DIRECT AND INDIRECT EFFECTS OF ACIDIC DEPOSITION ON VEGETATION

ACID PRECIPITATION SERIES
John I. Teasley, Series Editor

DIRECT AND INDIRECT EFFECTS OF ACIDIC DEPOSITION ON VEGETATION

Edited by Rick A. Linthurst

ACID PRECIPITATION SERIES—Volume 5
John I. Teasley, Series Editor

BUTTERWORTH PUBLISHERS
Boston · London
Sydney · Wellington · Durban · Toronto

An Ann Arbor Science Book

Ann Arbor Science is an imprint of Butterworth Publishers.

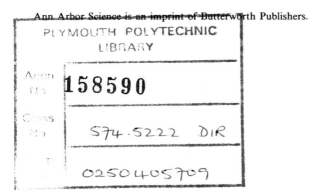

Library of Congress Cataloging in Publication Data
Main entry under title:

Direct and indirect effects of acidic deposition on vegetation.

(Acid precipitation series; v. 5)
Most of these papers were presented at the ACS Acid Rain Symposium, held in Las Vegas, Nev., March 28, 1982.
"An Ann Arbor Science book."
Includes index.
1. Acid precipitation (Meteorology)—Environmental aspects—Ad-
dresses, essays, lectures. 2. Plants, Effect of acid precipitation on—
Addresses, essays, lectures. 3. Acid deposition—Environmental as-
pects—Addresses, essays, lectures. 4. Forest ecology—Addresses,
essays, lectures. I. Linthurst, Rick A. II. ACS Acid Rain Sym-
posium (1982: Las Vegas, Nev.) III. American Chemical Society.
IV. Series.
QH545.Al7D56 1983   581.5'2642   83–18880
ISBN 0–250–40570–9

Butterworth Publishers
80 Montvale Avenue
Stoneham, MA 02180

10 9 8 7 6 5 4 3 2 1

Printed in the United States of America

CONTENTS

THE EDITORS

JOHN I. TEASLEY, SERIES EDITOR

John I. Teasley was a career employee of the U.S. EPA and several of its predecessor organizations. His background, both academically and throughout his professional employment, was that of an analytical chemist. His duties included research functions as well as serving in line management.

He was quite extensively involved in EPA's acid precipitation program from its inception until his retirement in July 1983. He continues to remain active in the ACS's Division of Environmental Chemistry and currently serves as the Councilor for the Lake Superior Section.

RICK A. LINTHURST, VOLUME EDITOR

Dr. Rick A. Linthurst is presently Director of Ecological Services for Kilkelly Environmental Associates in Raleigh, North Carolina. He was Program Coordinator for the North Carolina State University Acid Deposition Program for four years. He has a Bachelor of Science degree in the Biological Sciences, a Master of Science degree in Ecology, and a Ph.D. in Botany. He has been active in the field of acidic deposition effects on aquatic and terrestrial ecosystems since 1979.

Dr. Linthurst is a member of numerous scientific organizations and has received international recognition for his contributions to the sciences. He has served as a reviewer for ecological research programs established by many federal agencies, including the Environmental Protection Agency, the Department of Energy, and the National Space and Aeronautics Administration. Dr. Linthurst has focused his research activities on the interaction of plants and their environment, and ecosystem response to pollutant influences. Most recently, his research has emphasized regional assessments of precipitation quality, variability in precipitation chemistry, and integrative assessment as a tool for evaluating the causes, effects and cures for environmental problems.

THE CONTRIBUTORS

P.D. Brooks
Department of Plant and Soil Biology
University of California
Berkeley, California

Robert I. Bruck
Department of Plant Pathology
North Carolina State University
Raleigh, North Carolina

Christopher S. Cronan
Land and Water Resources Center
Department of Botany and Plant
   Pathology
University of Maine
Orono, Maine

M.K. Firestone
Department of Plant and Soil Biology
University of California
Berkeley, California

Jung Fuhrer
School of Forestry and Environmental
   Studies
Yale University
New Haven, Connecticut

Gordon Geballe
School of Forestry and Environmental
   Studies
Yale University
New Haven, Connecticut

Arthur H. Johnson
Department of Landscape
   Architecture and Regional Planning
Department of Geology
University of Pennsylvania
Philadelphia, Pennsylvania

K.S. Killham
Department of Plant and Soil Biology
University of California
Berkeley, California

Richard M. Klein
Botany Department
University of Vermont
Burlington, Vermont

Deborah G. Lord
Department of Geology
University of Pennsylvania
Philadelphia, Pennsylvania

Orie L. Loucks
The Institute of Ecology
Indianapolis, Indiana

J.G. McColl
Department of Plant and Soil Biology
University of California
Berkeley, California

Ellen T. Paparozzi
Department of Horticulture
University of Nebraska
Lincoln, Nebraska

Steven R. Shafer
Department of Plant Pathology
North Carolina State University
Raleigh, North Carolina

Thomas G. Siccama
School of Forestry and Environmental
    Studies
Yale University
New Haven, Connecticut

William H. Smith
School of Forestry and Environmental
    Studies
Yale University
New Haven, Connecticut

H.B. Tukey, Jr.
Urban Horticulture
University of Washington
Seattle, Washington

Robert S. Turner
Department of Geology
University of Pennsylvania
Philadelphia, Pennsylvania

SERIES PREFACE

These volumes are a result of a symposium on Acid Precipitation held in conjunction with the American Chemical Society's Las Vegas meeting, held in the Spring of 1982. The symposium was organized along nine thematic areas including meteorology, chemistry of particles, fogs and rain, oxidation of SO_2, deposition both wet and dry, terrestrial effects, aquatic effects, geochemistry of acid rain, economics, and predictive modeling.

The thematic areas were planned and conducted by expert investigators in each of the particular disciplines. The investigators were Chandrakant Bhumralker, meteorology; Jack Durham, chemistry of particles, fogs and rain; Jack Calvert, oxidation of SO_2; Bruce Hicks, deposition; Rick Linthurst, terrestrial effects; George Hendrey, aquatic effects; Owen Bricker, geochemistry; Thomas Crocker, economics; and Jerald Schnoor, modeling.

The symposium was designed to present findings on atmospheric movement of the precursors of acid precipitation; atmospheric chemistry and precipitation effects including aquatic, terrestrial and geochemical; the economics involved with "acid rain" and finally mathematical modeling in order that future problems may be predicted.

As one studies this series it becomes readily evident that the objectives of the symposium were indeed met and that the series will serve as a ready reference in the field of acid precipitation.

I wish to acknowledge the efforts and dedication of the volume editors for the superb job they did in bringing this series to fruition. Working with people such as these made all the time and energy spent well worth while.

I further wish to thank the American Chemical Society, The Division of Environmental Chemistry and the Council Committee of Environmental Improvement for providing the facilities for the symposium from which this series evolved.

Finally, I wish to express my gratitude to the individual chapter authors without whom the series would never have been possible.

John I. Teasley

PREFACE TO VOLUME 5

Acidic deposition is an extremely complex phenomenon. Speculation, fact and sound hypotheses have not always been clearly differentiated in the literature. As a result, many individuals outside the main stream of acidic deposition research find it difficult to accurately assess the issues. This is not an unexpected situation when scientists that study complex phenomena are asked to simplify, generalize and/or summarize their results for a variety of audiences.

The effects of acidic deposition on vegetation are as complex as the acidic deposition phenomenon itself. To date, few studies exist that support beliefs that long-term acidic deposition will negatively impact plant productivity. However, logic dictates that two plants, or two plant systems, exposed to different levels of sulfur, nitrogen and/or hydrogen ion will differ in their response. Because plants do not grow in isolation from other ecosystem components, e.g., differing soils, climate and air quality (the plant system as a whole), defining cause-and-effect relationships between acidic deposition and plant response increases in difficulty.

Compared to forest vegetation studies, considerable research has been completed on the effects of acidic deposition on crop plants. Both beneficial and detrimental impacts on crop species exposed to simulated acidic deposition have been observed, primarily in studies of soybeans. Some of the differences and results can be attributed to variability in experimental design, ambient air quality, soil characteristics and characteristics of the plant varieties used. However, because of the economic importance of crops, crop scientists in the United States have not ignored this differential response and are actively seeking a resolution to the controversies. Because of their "stand," homogeneity and short life-cycle, crops have served as "simple" systems on which one can examine the effects of acidic deposition on vegetation.

Research on the effects of acidic deposition on forest productivity has intensified over the past few years. Current data are not sufficiently convincing to support the hypothesis that acidic deposition has affected the growth of any economically important tree species. However, these data are sparse and a major effort is underway to determine if changes in forest productivity have occurred over the past 50 years. In this effort, the influence of acidic deposition will have to be distinguished from other anthropogenic and natural environmental stresses. Recent evidence suggests that the structure and function of high elevation forest communities have been altered. Acidic deposition and "acidic fog" may, in part, be responsible for these observations.

Potential impacts of acidic deposition on both crops and forests are very difficult to assess because a large number of environmental factors interact to determine plant productivity. In experiments focusing solely on acidic precipitation effects, plant responses have been extremely variable, ranging from inhibition of growth to no apparent effects to stimulation of growth. At the organism level, some damage, e.g., tissue lesions, has been demonstrated in controlled experiments, but the significance of these changes to the survival, reproduction and productivity of plants needs to be determined. Indirect effects of acidic deposition on plants may ultimately prove to be more significant than those expected from direct effects. The relationship between soil chemistry, nutrient cycling and the plant system may prove to be particularly enlightening areas of research as scientists search to define and understand the effects of acid inputs to terrestrial systems. Physiological and phenological changes in forest communities are also likely to play a role in long-term assessments. The knowledge gained from taking a holistic approach to the research, combined with an understanding of direct effects, will yield the most comprehensible understanding of the beneficial and/or detrimental effects of acidic deposition.

The eight chapters in this volume deal, in part, with the complexity of the system and the difficulty in defining cause-and-effect relationships between the plant, the plant system, and acidic deposition. The book begins with a presentation by Richard Klein, who provides a general overview of the problems that face all of us as we enter the acidic deposition effects arena from other fields of science. Dr. Klein has chosen to present an informal review of his research activities and leads the reader through the complex maze of ecosystem components which needs to be considered in holistic studies. Ellen Paparozzi and H.B. Tukey return from the complex ecosystem level to discuss direct effects of simulated acidic deposition at the plant level and characterize leaf injury following acid exposures. Robert Bruck and Steven Shafer consider effects of acidic deposition on plant diseases. This chapter is followed by an examination of potential interactions between acidic deposition, forest vegetation and microbially induced stress, presented by William Smith, Gordon Geballe and Jurg Fuhrer. Mary Firestone, John McColl, Kenneth Killham and Paul Brooks follow with an overview of the effects of acidic deposition on microbial components of the ecosystem and the subsequent effects on plant productivity. All three of the latter chapters are reminders that plants may not only be affected by acid inputs but that other ecosystem components might also be modified and as a result, indirect effects become increasingly important.

The three final chapters develop thoughts on the effects of acidic deposition at the forest community level. Christopher Cronan begins this series with a presentation of acidic deposition effects on chemical responses of forest canopies. Arthur Johnson, Thomas Siccama, Robert Turner and Deborah Lord discuss observed changes in forest tree growth rates in the northeastern United States and the evidence for linking these changes to acidic deposition. These chapters are followed by a conceptual presentation by Orie Loucks that highlights how forest site indices might be useful in defining the forest resources at risk from continued inputs of acidic substances.

The chapters in this volume only scratch the surface of the many factors that need to be examined relative to changes in plant systems that may result from atmospheric acidic inputs. This diversity of presentations has been selected to provide the reader with an overview of the variety of topics that are of primary concern to the research community. These presentations range from the general to the specific, from direct to indirect effects, and from a simple to complex consideration of all the interactive components. The authors represented by these contributions are active in the acidic deposition research arena, and the reader is encouraged to follow their progress in the future.

I would like to express my gratitude to the American Chemical Society for providing a forum to discuss vegetation/acidic deposition effects issues at its annual meeting in Las Vegas, Nevada, March 1982, where most of these papers were presented. Dr. John Teasley, who was instrumental in arranging the symposium dealing with acidic precipitation at that meeting, is deserving of special recognition. The authors who made this volume possible and the reviewers who helped enhance the scientific merit of each contribution are greatly acknowledged. Steven Vozzo and William Alsop of North Carolina State University are also acknowledged for their assistance in the review process and for "inspiring" the authors and reviewers to provide their materials on a timely basis. Finally, Ann Arbor Science is acknowledged for patiently waiting for the editor to provide his materials so that this volume could be printed.

Rick A. Linthurst

CHAPTER 1

Ecosystems Approach to the Acid Rain Problem

Richard M. Klein

When, as a plant physiologist, I started research on acid rain, I found myself almost overwhelmed by the conceptual and methodological difficulties involved in undertaking an ecological problem. As one who had spent his scientific life measuring to the third decimal place, it was difficult to accept the fact that two areas on the forest floor separated by 10 cm could be entirely different, and I now appreciate Thomas Brock's statement that "ecology is plant physiology under the worst possible circumstances." The diversity of subsystems that must be considered and the causal relationships among them proved to be difficult to fit together into a system that would allow the problem to be formulated. Cause and effect are not directly related, and the totality is certainly different from the sum of its parts. It was vitally important to erect a frame of reference on which to arrange and integrate the various facets of the study. Equally importantly, the conceptual scaffolding had to provide a heuristic tool that could point up specific parameters that required investigation in both laboratory and field. Out of necessity, we moved to ecosystems models.

There are many such models of varying complexity and sophistication. Those of Caswell et al. [1], Odum [2], Slobodkin et al. [3] and Luxmoor [4] received primary attention because their theoretical constructs and conceptual foundations lend themselves to studying the acid rain problem and because they facilitated the integration of other ecological studies that must be considered. Distilled from these models is an even simpler conceptual model (Figure 1). Essentially, it says that the anabolic and catabolic reactions of an ecosystem are cyclically linked, with plant growth and development (biomass accumulation) being dependent on the nutrients provided by the decomposition of debris (litter) and exogenous sources.

Each of the three major compartments of this model have, of course, impinging parameters that may also interact with each other. To use this model in our studies on acid rain, we took each of the major compartments and attempted to evaluate how acid rain could affect biomass accumulation, litter and nutrient pools. Our formulation, neither complete nor comprehensive, can be seen in Figures 2 to 4.

1

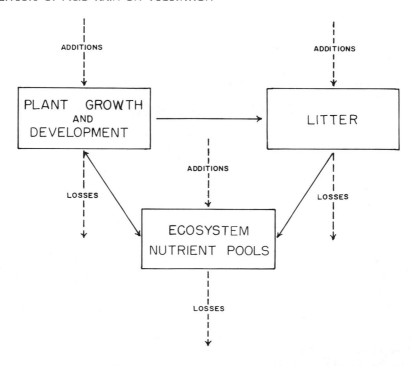

Figure 1. Generalized ecosystems diagram of anabolic and catabolic flows among major compartments of terrestrial ecosystems.

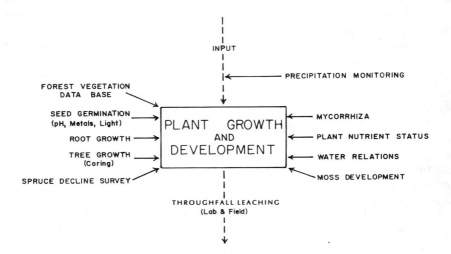

Figure 2. Diagrammatic representation of some major research topics that relate to the possible effects of acid rain on the growth and development compartment of terrestrial ecosystems.

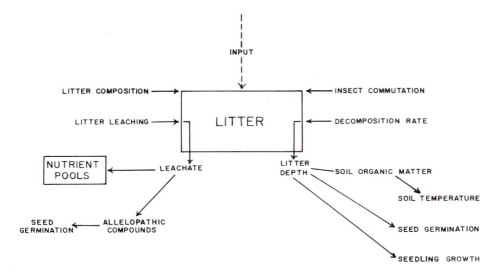

Figure 3.   Diagrammatic representation of some major research topics that relate to the possible effects of acid rain on the litter compartment of terrestrial ecosystems.

There are several caveats about our modeling. Each parameter can be considered separately (a reductionist approach) and the interactive parts can be reassembled (the synthesist approach). Since the thrust of our research is to determine whether or not acid rain is stressing or altering terrestrial ecosystems, the model allows us to ask if the process under study, e.g., litter leaching (Figure 3), can be altered by acidic precipitation, how the process is altered and whether the alteration is of sufficient magnitude to be a significant perturbation of the terrestrial ecosystem under study [5]. These evaluations are fraught with serious difficulties. Ecosystems are complex, resilient, holocoenotic, homeostatic and heterogeneous. Interrelationships among components are not always obvious and are frequently subtle. Thus, a demonstration that a perturbation of one facet of an ecosystem by acid rain is affecting one aspect of the dynamics of the system does not necessarily prove that the modulation resulting from this stress is significantly altering a major, or even a minor, step or process in the ecosystem as a whole. Ecosystem survival under difficult circumstances is well known.

Another caveat relates to proof. Assuming, without unequivocal evidence, that geographically extended acidic precipitation started in North America about 1950 or so, a substantial data base extending back to this time is simply not available. Our data base extends back to the mid 1960s and is almost unique in North America, but it is far from adequate. An equally serious limitation is the absence of an adequate control, if in fact a control in the classical sense is possible. Certainly no area in the northeast United States and adjacent Canada is currently without acid rain, and we are limited in developing research programs and are

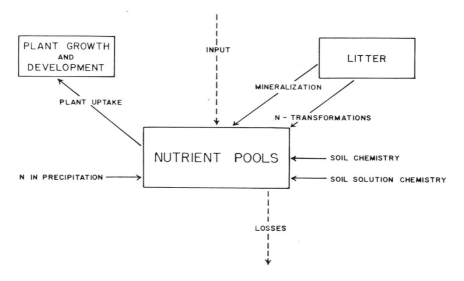

Figure 4. Diagrammatic representation of some major research topics that relate to the possible effects of acid rain on the nutrient pool compartment of terrestrial ecosystems.

forced to reach conclusions on the basis of incomplete field data. Whether control areas exist in the southern hemisphere may be worth exploring.

The conceptual ecosystem model that we have developed must be evaluated primarily on whether or not it is accurate in the context of the acid rain problem. I would like to use our own research program as a test of the model. Since we have been actively studying acid rain only since 1979, much of it is still in progress.

Most of our field work has focused on Camels Hump Mountain in the Green Mountains of Vermont. Camels Hump is one of the high peaks in New England, reaching 1240 m. It includes a northern hardwood forest from 550 to 730 m, a transition zone from 730 to 850 m, a boreal forest dominated by balsam fir and red spruce from 850 to 1160 m and an alpine zone to 1240 m. Although the hardwood and transition zones were logged, the boreal zone remains a virgin forest. Our field data base was provided by a doctoral thesis from the mid-1960s by T. Siccama [6] who did a thorough study of the ecology of the mountain. In 1979, the study was quantitatively repeated, the data computerized and compared for all floristic parameters. Although not all of the comparisons have been made, it was found that red spruce had declined significantly between 1965 and 1979 (Table I) [7]. Diminished growth rates and general low vigor plus considerable death and windfall of red spruce have been noted in 1981 and 1982, which indicates that the red spruce decline is continuing and possibly accelerating. As documented later in this report, there have been other changes of significance.

Along with others, we have found that the average pH of rain is close to 4.0 and that this rain contains a variety of heavy metals that appear to have

Table I.   Percentage Changes in Basal Area and Density of Dominant Tree Species on Camels Hump Mountain in Vermont Between 1965 and 1979[a]

Growth Zones	Species	Basal Areas (%)	Densities DBH[b] (cm)	%
Hardwood (550–730 m)	Sugar maple	−16		−20
	Beech	−30		−39
	Yellow birch	−29		+22
	Ash	−100		−100
	Red spruce	−40		−15
Transition (730–850 m)	Red spruce	−41		
Boreal (850–1160 m)	Balsam fir	−8	<2.0	+158
			2.0–9.9	+9
			10.0+	+21
	Red spruce	−45	<2.0	−32
			2.0–9.9	−28
			10.0+	−52
	White birch	−22	<2.0	−32
			2.0–9.9	+262
			10.0+	0

[a] Adapted from Siccama et al. [7].
[b] Diameter at breast height.

accumulated in the mountain soils. We have, in addition, been monitoring the fogs that condense on leaves and needles of trees in the boreal zone. On the average, the pH of this cloud water is close to 3.7, the ratio of sulfuric to nitric acid is different from that of rain, and the cloud water accounts for more than 50% of the total water reaching the boreal forest. Snow has the composition of rain, and rime ice has the composition of fog. Metal compositions of these forms of precipitation are being determined.

We have, then, two sets of facts. There are time-related alterations in forest composition and there is a significant deposition of acids and metals onto the plants and soils. It is necessary to determine whether there is a cause-and-effect relationship between these two sets of facts, and we have chosen to develop a program of simulation experiments to evaluate this possibility. The laboratory can provide information on whether a given stress can alter components of ecosystems, but it cannot determine whether the stress does alter the component in the forest. We are attempting to move the information obtained in the laboratory back into the field for such evaluations. At this time, we can make only probability statements and, through the integration of laboratory and field, obtain information that might allow predictions by inference and extrapolation. Quantitative balance sheets, such as those prepared by Likens et al. [8], can be invaluable aids in such deductive and inductive reasoning, but the current status of acid rain research does not permit this.

STUDIES ON PLANT GROWTH AND DEVELOPMENT

Individual research projects in the plant growth and development compartment of our model include our monitoring of precipitation and the floristic and ecological surveys noted above. Since it was found that spruce is most severely impacted, we have conducted studies on the effects of acidity and metals (Al, Cd, Cu, Pb and Zn) on germination of red spruce. At present precipitation pH, acid composition, and ionic metal concentrations, spruce and yellow birch germination was not affected. Light is a factor in seed germination, with significant repressions of red spruce germination by far red, green and blue wavelength bands. The relation of this finding to germination under forest conditions, where the canopy tends to filter out red and blue wavelengths, leaving enriched green and far red wavelengths, is under study.

Studies on the growth of spruce and balsam fir are still in progress. There may be a decline in growth as measured by widths of annual rings, but our original sample size was modest and our data were not subjected to the rigorous statistical analyses that are required. This study is being expanded. We are also interested in determining whether there was a time-dependent correlation between the possible decline in secondary growth of trees and metal accumulation in the xylem. Again, only preliminary data are available, but they suggest that metals, particularly Al, increased in the wood starting about 1950, the approximate time when acid rain may have begun in earnest. Studies on the relationship of metals and acids to cambial activity in vitro are under development.

In our field work, we were struck by our consistant failure to observe mycorrhiza on the roots of red spruce from the boreal zone, although we did find it on spruce from the hardwood zone. Microscopic studies confirmed these negative macroscopic observations. Since mycorrhizae are believed to be essential for the continued vigor of coniferous species, laboratory investigations of the effects of acidic conditions and metal ions on the growth of *Cenococcum graniforme* and *Polyporus circinatus,* two mycorrhizal fungi for spruce, were conducted. Al, Cu, Pb and Zn, presented in concentrations known or suspected to be present in boreal forest soils, repressed fungi growth in vitro, with greater repressions at pH 3.0–4.5 (that of the soils and soil water of Camels Hump) than at higher pH values. Studies on the effects of acidity and metal ions on the formation of the mycorrhizal association have yet to be done, but in view of the symptoms of spruce decline, the repressions of growth of the fungi appear to be of some importance.

If one follows the progression of spruce decline over a period of several years, one is struck by the observable sequence of changes. Essentially, the death of needles and the appearance of the trees is strikingly similar to that seen in cut Christmas trees between New Year's Day and Twelfth Night (January 6). This suggested that at least one of the complex of causal factors might be reduction in water uptake. Using seedlings of red spruce, we have investigated the loss of water by total transpiration (root uptake through stomatal diffusion) under controlled conditions of temperature, light and relative humidity. Experimental pH values of either 3.0 or 4.5 were not different, while the presence of Al and Cd in concentrations found in lysimeter collections significantly reduced transpiration. Pb and Zn were not repressive. Lysimeter-collected soil water also reduced total transpiration when early spring collections (pH 3.1) were tested, but summer collections (pH 3.7–4.2) were ineffective. It should be noted that the movement of water through trees is most active in the spring.

We, and others in Europe and North America, have found that there are alterations in the Al/Ca ratios in soils receiving acid rain and it has been reported that "fine roots" show degenerative changes under specific ratios. Laboratory simulations of root development of plants grown in hydroponic solutions with selected ratios of Al and Ca are underway.

As part of our floristic survey, the area of forest floor covered by mosses (not separated by species) was reduced in all three zones by close to 50%. Using a species that is dominant in all three zones, *Polytrichum ohioensis,* the growth capacity in vitro under controlled conditions was studied. Again, as with mycorrhizal fungi, the interaction of low pH and metal ions was pronounced. pH levels between 3.0 and 4.5 reduced vegetative growth of gametophytic plants by over 50% and the reductions were most severe in the presence of Al, Cu, Pb or Zn in concentrations known to be present in mountain soils. This does not prove that acidic, metal-containing precipitation caused the reduction in moss coverage, but it is consistent with the observed decline and is also consistent with other studies in severely impacted areas downwind from smelters.

Incoming precipitation in forested areas is largely filtered through the tree canopy before reaching the ground. In laboratory simulations using red spruce, we found that acidic precipitation at pH 4.3 leached more carbohydrate, protein, potassium and nitrate ion over three mistings separated by 72-h periods than did mistings with water at pH 5.6. These laboratory studies have been supplemented with field studies of throughfall with similar, but not identical results. It was found that throughfall under a red spruce stand was enriched in nitrate, sulfate, calcium and potassium, and that hydrogen ions were sorbed by the plants. Studies are underway at the laboratory level to confirm and extend these findings and to relate them to our soil water studies. Although chemical analyses of needles from presumably (visually determined) healthy and declining spruce trees by other workers suggested reduced potassium levels and a few other disparities in nutritional status, we are conducting simulations to determine whether the leaching in throughfall by acidic precipitation might have affected the mineral status of the needles. This study is, we believe, particularly important in view of our indications that water and, hence, nutrient movement may be impacted by reduction in transpiration and the absence of mycorrhizal associations.

STUDIES ON LITTER

Between 1965 and 1982, the depth of litter in all three floristic zones of Camels Hump increased by a factor of 2, a finding that is consistent with the reported reductions in litter decomposition in areas downwind from smelters plus a few other studies in Europe. Such reductions in litter decomposition and consequent reductions in mineralization of plant-required nutrients can have serious effects on forest plants.

We have simulated several facets of the litter and its nutrient relationships. Simulations of the effects of litter leaching by acidic and nonacidic snowmelt, fog, drizzle and rain showed that hydrogen ions from precipitation at pH 3 and 4 are rapidly sorbed by mixed balsam fir–red spruce litter. Conductivity of leachates was proportional to initial precipitation acidity, but the efflux of Ca, K and PO_4 ions was not greatly affected by pH.

Acidities of pH 3 and 4 repressed the respiratory metabolism of litter microorganisms, a measure of the initial stages in litter decomposition. Al and Cu, but not Pb and Zn, further depressed this activity. Although the percentage of organic matter in Camels Hump soils increased between 1965 and 1979, particularly in the boreal zone, a simulation of cellulose degradation (the conversion of cellulose to humus) failed to implicate either acids or metals. It is possible that the sorption of acid by litter is one of the factors depressing microbially caused litter decomposition.

Microorganisms are not the only biotic system involved in litter breakdown; several kinds of commutating invertebrates are actively involved in the process. Using a data base from the early 1970s, we are investigating the species distribution and numbers of Collembola that we have isolated from the litter of all three floristic zones. At this time, useful data have not been obtained, although, based on several older studies in Europe, there is reason to suspect that these organisms may be affected by acidity.

The increased depth of litter might have other effects on the forests. We have not been able to find seedling spruce in the boreal zone of Camels Hump. As noted above, acidic, metal-containing precipitation does not, under laboratory conditions, affect spruce germination, and we have found some cones containing apparently viable seed. Under laboratory conditions, we found that seeds of red spruce, balsam fir and yellow birch failed to develop into seedlings when sown either on top of or beneath depths of balsam fir–spruce litter exceeding 0.5 cm. Further laboratory studies demonstrated that this litter contained one or more allelopathic substances affecting red spruce and yellow birch, and that these substances could be leached out with either acidic (pH 4) or nonacidic (pH 5.6) water. Since considerable litterfall occurs in the autumn and most seed germination occurs in the spring after vernalization (stratification) of the seed, it seems unlikely that the presence of allelopathic litter is a major factor in our failure to find spruce seedlings. There is only limited information on the smothering of seedlings by litter, and this topic is under investigation in our laboratory. It is, however, of more than passing interest that the coniferous litter is highly allelopathic to yellow birch germination, since this plant is rarely found in New England associations dominated by these conifers.

STUDIES ON NUTRIENT POOLS

At this time, there is inadequate evidence that there are reductions in the levels and ratios of inorganic nutrient ions available to the flora on Camels Hump. Our studies on litter suggest that there may be less available, since litter decomposition has been reduced and organic matter—which can bind many ions—has increased. Two general approaches to this important matter are immediately apparent. The soil chemistry can be investigated, as can the chemistry of soil water. Both approaches are being used by our group. Fortunately, we retained the soil samples collected in 1965 and are comparing them with soils collected now. These analyses are still in progress. Suction lysimeters have been installed at numerous locations on Camels Hump and the soil water collections made at weekly intervals are also being analyzed.

In forest soils, both nitrogen and phosphorus tend to be in limiting supplies. We have focused our attention on nitrogen metabolism of the Camels Hump forests. Large soil cores have been taken from all ecosystems and are being percolated with acidic and nonacidic precipitation to measure release of ammonium and nitrate nitrogen over several months. This is to be followed by studies on the nitrification potential of these mountain soils and, possibly, by assays for nitrogen-transforming microorganism activity.

There has been some interest in whether the nitrogen compounds in precipitation are of significance for plant nutrition. An evaluation of data on the chemical composition of rain extending back to the middle of the nineteenth century in England and to the turn of the century in the northeast United States has been done. These literature reports, plus more erratic and incomplete records for other locations in the northern hemisphere, have been evaluated after conversion of the raw data to a standard base. We found that the ammonium/nitrate nitrogen ratio was close to 3 for a century, after which the ratio fell to 0.5 when, in about 1950, rains began to increase in acidity with nitric and sulfuric acids dominating. Statistical treatments, still not complete, suggest that the ammonium nitrogen levels may not have been altered drastically during the time period of the study. On the average, and with due allowance for changes in analytical procedures, meteorological inconsistencies and sample sizes, there are about 4–6 kg-ha^{-1}-y^{-1} of soluble nitrogen reaching the temperate land and its plants. This amount is of little importance for agronomic crops, since they require amounts exceeding 100 kg-ha^{-1}-y^{-1}. Whether precipitation nitrogen is of importance for forested ecosystems is still not known, and an assessment of this possibility will depend on information still to be obtained from studies on soil water, soils and analyses of plants. Since the release of nutrients from litter is likely to be reduced, it may be important.

Studies on nitrogen fixation in forested lands is still not complete. Camels Hump, like most New England mountains, does not have plants possessing associative nitrogen-fixers. We, and others, have isolated free-living *Azotobacter*, but our laboratory studies have shown that the acid levels of Camels Hump soils fully repress their growth and there are many reports that they do not fix nitrogen at these acid pH values. Blue-green algae could not be isolated from soils, tree bark or leaves on Camels Hump, and laboratory studies demonstrated that axenic cultures die at pH values below 4.0 and that, even at pH 5 or 6, the presence of metal ions was lethal. To our surprise, we have failed to find anaerobic nitrogen-fixing clostridia in soils from Camels Hump, although nonfixing species have been found. We have, however, isolated free-living, N-fixing, aerobic microorganisms from all ecosystems of Camels Hump that grow well under highly acidic conditions. Their taxonomy and metabolism are under study.

DISCUSSION

It is obvious that there are many alterations in the composition and development of plants on Camels Hump Mountain and that simulation studies show that acidic, metal-containing precipitations are capable of causing alterations in the growth and metabolism of components of these ecosystems. It is difficult to determine how far we can extrapolate the laboratory data to relate it to the data obtained in the field. The decline and death of red spruce, the reduction in moss coverage, the increase in litter depth, the accumulation and solubilization of metals in the soils, the absence of mycorrhiza on spruce roots and other alterations noted since the initial ecological study suggest strongly that the forests of Camels Hump (and several other surveyed mountains in Vermont) are reacting to severe stresses. Our studies indicate that disease or insect infestations are not involved, and we do not believe that the changes we have observed and measured are due to the natural ebb and flow of ecosystem life. We believe acid rain is a reasonable candidate as a major perturbing influence, but the evidence from our work and the work of others is still inadequate to make a firm probability statement. The available data, both laboratory and field, are not sufficient to allow us to assert with any confidence how much these natural ecosystems are being altered, where and how the alterations are occurring, and whether the alterations are of sufficient magnitude to be a "clear and present danger."

However, it must be remembered that any ecosystem is a delicately balanced interaction among a very large number of biotic and abiotic components. It is virtually an ecological law that if even one component of an ecosystem is altered, perturbed or stressed, the direct and indirect consequences will be perceived and responded to by all other components and that each of these, in turn, will further shake the web of life. Whether or not acidic, metal-containing precipitation is the cause of spruce decline and other alterations, anthropogenically altered precipitation cannot fail to stress natural ecosystems. We must also remember that reductions in complexity of ecosystems by loss or pauperization of biota will reduce the stability of that ecosystem.

As an article of faith, I believe that the matter of acid rain, which is of great ecological, environmental, social, economic and political importance, can be understood by combined field and laboratory investigation.

ACKNOWLEDGMENTS

Interactions with Frank Reed and H. W. Vogelmann are acknowledged with appreciation. This is journal paper 510, Vermont Agricultural Experimental Station, supported by the Vermont Agricultural Experimental Station under the Hatch Act, the Northeast Forest Experimental Station and the American Electric Power Service Corporation.

REFERENCES

1. Caswell, H., H.E. Koenig, J.A. Rech and Q.E. Ross. "An Introduction to Systems Science for Ecologists," in *Systems Analysis and Simulations in Ecology, Vol. II*, B.C. Patten, Ed. (New York: Academic Press, Inc., 1972), pp. 3–78.
2. Odum, H.T. "An Energy Circuit Language for Ecologists and Social Systems: Its Physical Basis," in *Systems Analysis and Simulations in Ecology, Vol. II*, B.C. Patten, Ed. (New York: Academic Press, Inc., 1972), pp. 139–211.
3. Slobodkin, L.B., D.B. Bodkin, B. Maguire, Jr., B. Moore III and H. Morowitz. "On the Epistemology of Ecosystem Analysis," in *Estaurine Perspectives* (New York: Academic Press, Inc., 1980), pp. 497–507.
4. Luxmoore, R.J. "Modeling Pollutant Uptake and Effects on the Soil-Plant Litter System," in *Effects of Air Pollutants on Mediterranean and Temperate Forest Ecosystems*, P.R. Miller, Ed. (Berkeley, CA: Pacific Southwest Forest and Range Experiment Station, 1980), pp. 174–180.
5. Smith, W.H. *Air Pollution and Forests: Interactions Between Air Contaminants and Forest Ecosystems* (New York: Springer-Verlag, 1981).
6. Siccama, T. "Altitudinal Distribution of Forest Vegetation in Relation to Soil and Climate on the Slopes of the Green Mountains," PhD Thesis, University of Vermont (1968).
7. Siccama, T., M. Bliss and H.W. Vogelman. "Decline of Red Spruce in the Green Mountains of Vermont," *Bull. Torrey Bot. Club* 109:162–168 (1982).
8. Likens, G.E., F.H. Bormann, R.S. Pierce, J.S. Eaton and N.M. Johnson. *Biogeochemistry of a Forested Ecosystem* (New York: Springer-Verlag, 1977).

CHAPTER 2

Characterization of Injury to Birch and Bean Leaves by Simulated Acid Precipitation

Ellen T. Paparozzi
H.B. Tukey, Jr.

In the last 20 years, scientists in Western Europe, central Canada and the eastern United States have recorded decreasing pH levels in rainwater. This phenomenon has been labeled as "acid rain." Acid rain is any rain with a pH of less than 5.6, taking into account acidity contributed only by the concentration and partial pressure of CO_2 in the atmosphere. Thus, acid rain is actually a dilute solution of sulfuric and nitric acids rather than a neutral solution.

Most work involving acid rain and its interaction with plants has been performed in vitro, and the list of plants that are sensitive to simulated acid rain continues to grow [1]. In many cases experimenters have found that plants such as bean [2–6] and birch [7] will show injury symptoms, described as tan-to-yellow lesions, when plants are exposed to simulated acid rain of pH 3.2 or less.

Exactly how simulated acid rain enters into plants was explored by Shriner [3,4]. Using stereo scanning electron microscopy (SEM), Shriner viewed the surfaces of injured leaves of *Quercus phellos* and *Phaseolus vulgaris* cv. Red Kidney. He reported that these plants showed erosion of cuticle surface wax and pits corresponding to where blocks of cuticular wax should have been. Untreated leaves showed no erosion of wax. He then suggested that this cuticular erosion could promote acid rain injury to leaves.

This hypothesis seemed reasonable, as the cuticle and its epicuticular wax component are the primary barriers to penetration of substances that come in contact with the leaf surface. Thus, a more in-depth study could further verify this phenomenon.

To accomplish this, a leaf developmental study was performed by sampling birch (*Betula alleghaniensis* Britt.) leaves at one-week intervals for four weeks and bean (*Phaseolus vulgaris* cv. "Red Kidney") leaves, specifically the second trifoliate leaf, at four similar growth stages. Leaves were fixed for and examined by light microscopy (LM), SEM and transmission electron microscopy (TEM).

13

Through the use of these methods, surface characteristics (such as trichome type and density, and presence and structure of epicuticular wax) and internal characteristics (such as cuticle thickness and cell types) were determined. Leaf area and midvein length were also measured. On the basis of this study one leaf stage per plant was selected for exposure to simulated acid rain.

Many important factors, such as rain droplet size and terminal velocity, type and frequency of rain, and droplet dry-down time, were considered in this design [1]. Simulated rain of pH 2.8, 3.2, 4.3 or 5.5 was applied for 6 min followed by 6 min of drying time for a total of 2 h-d^{-1} for four consecutive days. Leaves of varying ages, but especially birch leaves 21 d old and bean leaves approximately 11–16 d old from plants exposed to simulated acid rain were fixed and studied using the same preparations and methods of microscopy as in the initial developmental study.

Halfway through the acid rain simulation experiments, leaves that received simulated acid rain of pH 2.8 showed signs of injury. On termination of these experiments, leaves that had been exposed to pH 3.2 simulated acid rain also showed signs of injury. The injury, as seen macroscopically on birch, appeared as small, round, yellow lesions with or without necrotic edges. Injury to bean leaves appeared as small tan lesions initially on the unifoliate leaves and later on the trifoliate leaves. In both cases, microscopic surface viewing (SEM) revealed lesions as collapsed epidermal cells with epicuticular wax still present. All trichome types except glandular trichomes on birch were uninjured, but often appeared associated with lesions. Figure 1A shows this injury (l-lesions) amid apparently uninjured epidermal cells on leaves of red kidney bean. Figure 1B shows the midrib of the bean leaf dividing uninjured cells on the right and injured cells on the left. Defensive and glandular trichomes were uninjured on either side.

When thin sections were cut through lesions and viewed with LM, injured areas appeared as collapsed or collapsing epidermal cells with plasmolyzed palisade cells. In Figure 1C, this injury to a birch leaf was associated with an unicellular trichome and a vascular bundle. When areas of extreme injury on birch leaves were viewed using TEM, epidermal cells appeared completely collapsed and cell contents of the palisade and epidermal cells were indistinguishable (Figure 2A). When the epidermal cell wall was magnified about 29,000X, as in Figure 2B, the cuticle was still present. No difference was found between cuticle thickness over injured cells vs uninjured [8] (Figure 2C).

These results show that when viewing the surface of a leaf that has been injured by simulated acid rain, the epicuticular wax is still present. When sections are cut through the leaf, the cuticle was found to be present with little reduction in its thickness. These results are in contrast with those of Shriner [3].

Thus, it appears that simulated acid precipitation is entering the leaf and causing damage after passing through rather than eroding the cuticle away. These results should not be surprising, since for most leaves the cuticle serves as a weak cation exchanger [9] and many substances in aqueous solution, such as pesticides and nutrients, are absorbed by foliage [10–14].

Although the exact steps in the process of cell injury could not be determined

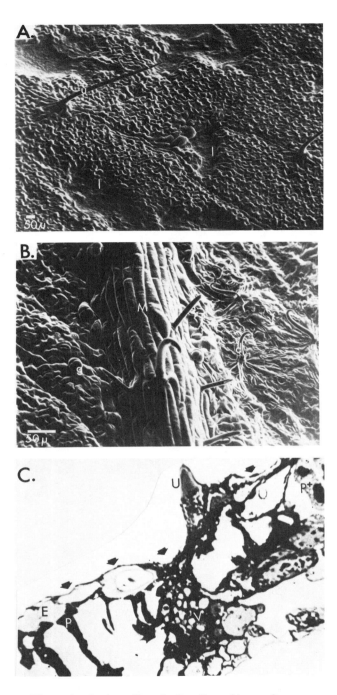

Figure 1. Lesions (l) on leaf surface of bean after exposure to simulated acid rain of pH 3.2 and 2.8 on A. lamina epidermis; (u) unicellular trichome B. midrib (M) and surrounding epidermis; (g) glandular trichomes and (d) defensive trichomes C. birch leaf injured by pH 2.8 acid rain (arrows); (E) epidermis, (U) unicellular trichome, (P) injured palisade cells, (P$^+$) uninjured palisade cells.

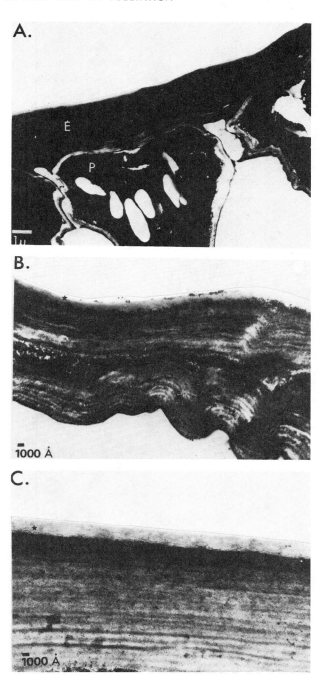

Figure 2. Thin sections of birch leaf showing A. injured epidermal (E) and palisade cells (P) B. cuticle (*) over injured lamina epidermal cell and C. over control lamina epidermal cell.

definitively, a pattern did appear. In most cases, epidermal cells would be injured first, often losing their contents but not necessarily collapsing. Then the palisade cells' contents became darkened and indistinct, with plasmolysis occurring simultaneously or soon thereafter. Work by Ferenbaugh [15] and Hindawi et al. [6] using *Phaseolus vulgaris* L. and *P. vulgaris* cv. Contender, respectively, showed a similar loss in chloroplast integrity and a decrease in chlorophyll content when plants were exposed to simulated acid rain of pH 3.0 and lower. In addition, Ferenbaugh found that the contents of injured palisade cells had a pH of less than 4.0. For comparison, hornwort has a cytoplast and vacuolar pH of approximately 6.6 [16].

Although it is known that the pH of cell contents may vary from plant to plant, every cell has a range of acidity that it can tolerate. If this range is exceeded, one or more of the processes that regulate cell function may be disrupted [16]. In the case of sulfuric acid, as in simulated acid rain (strong acid), on arrival near cells it will ionize to form sulfate and hydrogen ions. Thus, the cell wall could become acidic and a high ionic strength of ions would be formed on the outside of the cell, resulting in loosening of the wall fibers. It is possible that a high osmotic potential induced by sulfate in combination with cell wall acidification could cause a drop in turgor, as suggested by Heath [17]. This imbalance could lead to water loss from the cell, thus causing a loss of turgidity of epidermal and palisade cells plus eventual plasmolysis [18]. Thus, injury to epidermal and palisade cells could be due to a change in pH levels from the additional H^+ ions available in the acid rain.

Whether the injury symptoms produced by this simulated rain system will mimic plant injury in the field is uncertain. No attempt was made to simulate the increase in alkalinity (in terms of pH) that occurs during a natural rain shower as the steps involved in leaf injury have yet to be elucidated.

ACKNOWLEDGMENTS

This is paper No. 81–1639, journal series, Nebraska Agricultural Experiment Station. Research was conducted at Cornell University, Ithaca, New York.

REFERENCES

1. Paparozzi, E.T. "The Effects of Simulated Acid Precipitation on Leaves of *Betula alleghaniensis* Britt. and *Phaseolus vulgaris* cv. Red Kidney," PhD Thesis, Cornell University, Ithaca, NY (1981).
2. Wood, T., and F.H. Bormann. "Increase in Foliar Leaching Caused by Acidification of an Artificial Mist," *Ambio* 4:169 (1975).
3. Shriner, D.S. "Effects of Simulated Rain Acidified with Sulfuric Acid on Host-Parasite Interactions," PhD Thesis, North Carolina State University, Raleigh, NC (1974).
4. Shriner, D.S. "Effects of Simulated Rain Acidified with Sulfuric Acid on Host-Parasite Interactions," *Water, Air, Soil Poll.* 8:9–14 (1977).
5. Evans, L., N.F. Gmur and J.J. Kelsch. "Perturbations of Upper Leaf Surface Structures by Simulated Acid Rain," *Environ. Exp. Bot.* 17:145–149 (1977).

6. Hindawi, I.J., J.A. Rea and W.L. Griffis. "Response of Bush Bean Exposed to Acid Mist," *Am. J. Bot.* 67(2):168–172 (1980).

7. Wood, T., and F.H. Bormann. "The Effects of an Artificial Mist upon the Growth of *Betula alleghaniensis* Britt.," *Environ. Poll.* 7:259–268 (1974).

8. Paparozzi, E.T., and H.B. Tukey, Jr. "Developmental and Anatomical Changes in Leaves of *Betula alleghaniensis* Britt. and *Phaseolus vulgaris* cv. Red Kidney Before and After Exposure to Simulated Acid Precipitation," *J. Am. Hort. Sci.* (in press).

9. Kollatukudy, P.E. "Biopolyester Membranes of Plants: Cutin and Suberin," *Science* 28:990–1000 (1980).

10. Paparozzi, E.T., and H.B. Tukey, Jr. "Foliar Uptake of Nutrients by Selected Ornamental Plants," *J. Am. Soc. Hort. Sci.* 104(6):843–846 (1979).

11. Reed, D.W., and H.B. Tukey, Jr. "Effect of pH on Foliar Absorption of P and Rb Compounds by Chrysanthemum," *J. Am. Soc. Hort. Sci.* 103:337–349 (1978).

12. Hull, H.M., H.L. Morton and J.R. Wharrie. "Environmental Influences on Cuticle Development and Resultant Foliar Penetration," *Bot. Rev.* 41:421–452 (1975).

13. Hull, H.M. "Leaf Structure as Related to Absorption to Pesticides and Other Compounds," *Residue Rev.* 31:1–155 (1970).

14. Franke, W. "Mechanisms of Foliar Penetration," *Ann. Rev. Plant Physiol.* 18:281–300 (1967).

15. Ferenbaugh, R.W. "Effects of Simulated Acid Rain on *Phaseolus vulgaris* L. (Fabaceae)," *Am. J. Bot.* 63(3):283–288 (1976).

16. Smith, F.A., and J.A. Raven. "Intracellular pH and Its Regulation," *Ann. Rev. Plant Physiol.* 30:289–404 (1979).

17. Heath, R.L. "Initial Events in Injury to Plants by Air Pollutants," *Ann. Rev. Plant Physiol.* 31:395–491 (1980).

18. Curtis, O.F., and D.G. Clarke. *An Introduction to Plant Physiology* (New York: McGraw-Hill Book Company, 1950).

CHAPTER 3

Effects of Acid Precipitation on Plant Diseases

Robert I. Bruck
Steven R. Shafer

The chemistry of rain in much of the eastern United States has been modified to such an extent that levels of acidity considered insufficient to induce acute botanical effects may indeed significantly influence the epidemiology of numerous plant diseases. The role of acid rain on the subtle balance between host and parasite life cycles has been and continues to be a subject of scientific investigation. Evidence to date suggests that acid precipitation may significantly alter host–parasite interactions and hence modify otherwise predictable disease responses. Although the majority of available data were gathered under greenhouse or field "simulated rain" conditions, we present experimental evidence that addresses potential effects of acid rain on plant disease.

Many definitions for the concept of "disease" in plants have been proposed over a period of decades [1], but none is completely satisfactory to all. Like most "definitions" in the biological sciences, problems arise with the gray areas of those situations. Some would insist that the altered physiological processes induced by acid precipitation itself constitute bona fide "disease," and we would be hard-pressed to argue effectively against such a claim. However, to keep this chapter in its proper context with other topics in this volume, our concept of disease will be limited to processes in plants typified by injurious physiological activity induced by biotic pathogens (fungi, bacteria, nematodes, parasitic seed plants, mycoplasmas, spiroplasmas, rickettsias, protozoans, viruses and viroids) over a period of time. This idea can be presented in the form of the familiar "disease triangle" (Figure 1), which indicates that disease is a function of the simultaneous occurrence of a susceptible host plant, a virulent parasitic organism and environmental conditions conducive to their interaction over time. This environmental component is as important to disease development as the host or parasite. Without favorable environmental conditions, the disease process will not occur.

Acid precipitation represents an alteration in the environment component

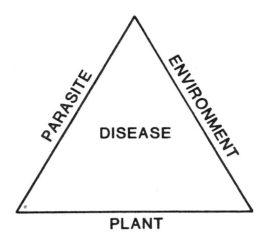

Figure 1. Disease triangle. Disease (as presented in this chapter) is a function of a physiological interaction between a plant host and a parasite in an environment conducive to that interaction, over a period of time. Acid precipitation represents a factor in the "environment" component. The amount of disease, in terms of incidence or severity, is reflected in the size of the triangle.

(represented by a star in Figures 2 and 3). Hypothetically, it could have three effects on a disease situation. First, within the range of acidity currently occurring in precipitation, no change may be evident. Either the perturbation is simply not sufficient to alter the otherwise predictable course of the host–parasite interaction, or it produces a balance of conflicting effects. Second, acid rain may influence host resistance, pathogen virulence or pathogen inoculum density in such a way as to increase disease incidence or severity on a plant or within a plant population (Figure 2). Third, acid precipitation may decrease plant susceptibility, parasite virulence or parasite inoculum such that the disease process is decreased in severity or incidence (Figure 3). Presumably, the ultimate effect will vary with the host–parasite situation under consideration, the range of rainfall acidity involved and the impact of acidity on each of the myriad phases of the host–parasite interaction.

The literature addressing this specific topic is scant. However, knowledge of the biology of plants, their parasites and the interactions of these populations in the environment can suggest ways in which acid rain might influence plant diseases. This type of speculation is valuable in directing research efforts into these topics, and we offer some of these hypotheses here.

This chapter is divided into two broad sections: (1) effects of acid rain on airborne plant diseases and foliar and stem pathogens; and (2) effects on root

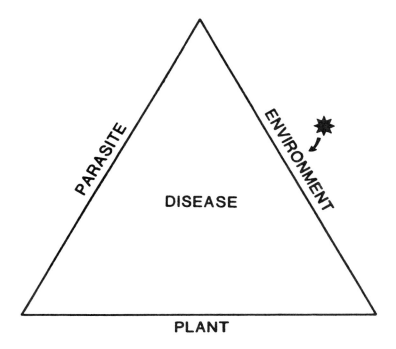

Figure 2. Acid precipitation (*) in the environment may influence some plant–parasite interactions in a manner that results in an increase in disease incidence or severity.

Figure 3. Acid precipitation (*) may influence some plant–parasite interactions such that overall disease incidence or severity is decreased.

diseases and soilborne pathogens. Hypothetical influences on beneficial root–microorganism symbioses will be mentioned briefly. Using the disease triangle as a guide, the effects of acidic precipitation on diseases will be approached through the perspective of plant-mediated and parasite-mediated effects.

EFFECTS OF ACIDIC PRECIPITATION ON FOLIAR AND STEM DISEASES AND ON FOLIAR-STEM PATHOGENS

Plant-Mediated Effects

Recent research has addressed direct effects of acid rain on the deterioration of plant tissue. Literature exists on the role of the cuticle, epicuticular waxes and the epidermis as barriers to penetration by foliar plant pathogens [2]. Presumably, when acid rain erodes leaf surfaces and kills tissue, the consequences for the microecology of leaf surfaces may be great. Changes in the leaf surface environment could favor or inhibit ingress by pathogenic microorganisms.

Researchers have speculated on the possible significance of direct weathering of leaf surfaces by acid rain, and such weathering may directly or indirectly influence plant—parasite interactions. Rain, and indeed, acid rain, is probably the most significant environmental factor influencing the weathering of plant tissue [3,4]. Martin and Juniper [4] observed that weathered leaf surfaces exhibited increased surface wettability, often a critical factor in pathogen propagule survival and ingress. They also noted that weathered surfaces may pose a weaker barrier to direct penetration of pathogens. However, weathered leaf surfaces may aid in retention of pesticides due to the erosion of hydrophilic epicuticular waxes that would otherwise shed residues [5].

Plant stress and wounds accelerate senescence of plant tissue. Infection by an entire group of plant pathogens, i.e., the facultative parasites, is favored by senescent tissue. Hence, the incidence and severity of plant diseases induced by many of the ascomycetous and basidiomycetous fungi that are placed in this category may be increased. Conversely, populations of obligate parasites, i.e., some oomycete fungi, rust fungi, powdery mildews and viruses that require healthy tissue for infection may be significantly reduced by acid rain stress on plants. For example, Bruck et al. [6] observed that when loblolly pine (*Pinus taeda* L.) seedlings were exposed twice before and after inoculation with basidiospores of *Cronartium quercuum* (Berk.) Miyabe ex Shirai f. sp. *fusiforme* (the fusiform rust pathogen) to episodes of acid rain at pH 5.6, 4.0, 3.2 and 2.4, needle necrosis ranged from 0% at pH 5.6 to 60% at pH 2.4. Six months following inoculation they noted significantly fewer rust galls among trees previously exposed to lower-pH rains. Basidiospore penetration takes place through healthy tissue of the needle or fascicle junction. Induction of tissue necrosis by acidified rains would reduce the potential target area; hence, fewer galls would be expressed (the result of lower infection rates).

Diseases caused by fungi that act as facultative saprophytes would be expected to develop differently from those induced by obligate parasites. Numerous fungi that cause tree cankers, i.e., *Nectria* spp., *Hypoxylon* spp., *Eutypella* spp. and others, require wounds to gain ingress. Preliminary reports [7] indicate that the increased acidification of rains in the northeastern United States may correlate with observed increases in stem cankers. It is difficult to separate these observations from the effects of other environmental perturbations, but the correlations warrant further investigation.

Stapp [8] notes that most bacterial plant pathogens gain ingress into plants via wounds and natural openings. Microscopic wounds induced by low-pH rains may act as courts for bacterial infection. Shriner [9] exposed bean leaves to simulated rains of pH 3.2 or 6.0 and noted a marked increase in the incidence and severity of lesions formed by the bacterial pathogen *Pseudomonas phaseolicola* (Burkh.) Dows at the lower pH. Since simulated acid rain clearly induces plant wounding, Shriner's hypothesis that simulated rains induced numerous small necrotic lesions, resulting in an increase in bacterial pathogenesis, seems to be supported.

Acid rain may have both detrimental and beneficial effects on host plant physiology with respect to response to foliar plant pathogens [9]. Acid precipitation may disturb otherwise normal carbon and nutrient metabolism; hence, defense mechanisms may be altered. However, plants have evolved highly efficient anatomical and biochemical processes to respond to both environmental perturbation and pathogen ingress [10]. The ability of a plant to defend or repair itself depends on the availability and allocation of resources. Acid rain can directly deprive a plant of necessary resources via foliar and soil leaching of nutrients and exudates, or it may indirectly shift the physiological balance of resources from defense to repair of injured tissue.

Many plants react to injury (as may occur in episodes of acid rain) by "shifting" metabolism into defensive-related pathways. Hypersensitive reactions that "trap" obligate parasites within dying tissues have been observed in numerous instances, such as in certain viral diseases and some rusts. Compartmentalization of fungal invasion in trees by the mobilization of antibiotic compounds has been documented. Bruck et al. [6] reported a significant increase in anatomical "resistant traits" in loblolly pine seedlings exposed to low-pH rain treatments both before and after inoculation with the fusiform rust pathogen. They speculate that the acidified rain insult may, by itself or in concert with fungal invasion, induce the host plants to contain or halt the process of *Cronartium* pathogenesis. The incidence of abnormally small galls was greater when the pH of simulated rains was lowered. Low pH also correlated to an increase in reaction zones (possibly accumulations of phenolics), galls with rough exfoliative bark tissue (correlated to field resistance) and infection sites with no gall symptom expression.

The potential ramifications of these plant-mediated effects of acidified rain to plant disease epidemiology are many. Pathogen propagules not only may apparently be killed or rendered nonviable, but they may also be contained following initial infection and colonization of plant tissue by the effects of acid rain. Data collected as of this writing suggest that rains of high acidity may significantly alter otherwise predictable plant-mediated disease responses in nature.

Parasite-Mediated Effects

Plant pathogens may directly be altered or destroyed by the effects of acid rain. Numerous species of fungi and bacteria are disseminated via wind-blown or splashed rain. The ability of many bacteria to survive in an acidified environment is poor [11]. In addition, Leben [12] observed severe growth reduction in 15 species of plant pathogenic fungi grown on media acidified to pH 3.2. Shriner [13] found that the acidity of water in which *Pseudomonas phaseolicola* bacterial inocula were disseminated onto bean greatly influenced both the bacterial population and infectivity. At pH 3.2, bacteria that remained in the infection droplets were incapable of infecting either healthy leaves or foliage injured by acid rain.

Lacy et al. [14] studied the effects of simulated acidified precipitation on populations of the bacteria *Erwinia herbicola* (Lohnis) Dye and *Pseudomonas syringae* V. Hall. They observed that no colony-forming units of these epiphytic bacteria were recovered on agar plates after exposure for 50 min to acidified water at pH 2.0 and that there was a significant depression of bacterial populations at pH 3.0, as compared to bacteria exposed to pH 4.0, 5.0, 6.0 or 7.0.

Bruck et al. [6] hypothesized that fusiform rust basidiospore inocula exposed to episodes of simulated acid rain following inoculation to loblolly pine may have been reduced in its infectivity. Highly significant differences in the number of galls initiated were noted among trees exposed before and after inoculation with the rust fungus. The authors postulated that the reduction of galls at low-pH exposures may have been due to the reduction of viable propagules available to initiate the infections. The probability of successful pathogen ingress would most probably be reduced as fewer viable propagules would initiate infections.

EFFECTS OF ACID RAIN ON ROOT DISEASES AND SOILBORNE PATHOGENS

Acid precipitation may affect diseases caused by soilborne pathogens by mechanisms similar to those influencing foliar diseases, i.e., in plant- or pathogen-mediated ways. Many of these mechanisms may be induced through acute effects of short-term acidification of soil water. In addition, the potential for long-term effects on both mechanisms due to chronic exposures is of possible significance.

Plant-Mediated Effects

Like other phases of plant growth and development, optimum seed germination occurs within a certain pH range. For example, Teigen [15] demonstrated that the greatest percent seed germination of Norway spruce seeds occurred at pH 4.8. Lower germination percentages were observed in soils drenched with dilute sulfuric acid or amended with lime. Furthermore, at pH 3.8, 80% of the seedlings that did emerge were not "normal." Exact response, whether positive or negative, of germinating seedlings to simulated acid rain is species-specific [16]. Plants for which early emergence and establishment are delayed are often susceptible to attack by damping-off pathogens over a longer period of time and thus have increased chances of being killed. Exposure of seeds germinating near the soil surface to acid solutions entering the soil as rainwater might induce such delays in emergence and early development.

Simulated acid rains can induce direct damage to plant foliage and, as discussed previously, such acute damage to tissues can alter plant susceptibility to pathogen attack. Root tissue is known to be sensitive to acidic solutions as well. Cohen et al. [17] noted significant reductions in root fresh and dry weights for carrot plants exposed throughout the growing season to simulated rain acidified with sulfuric acid to pH 4.0 or below, compared with yields for plants exposed to pH 5.6 treatments. Yields (fresh and dry weights) of beet roots were reduced by pH 3.0 exposures. Arnon and Johnson [18] demonstrated that lettuce and tomato plant roots turned gray, lost turgor and collapsed when exposed to nutrient solutions adjusted to pH 3.0. Similar effects by low-pH rain on roots near the soil surface could lead to the presence of senescent tissue that might provide infection courts for facultative parasites in soil. Such an effect may have been observed by Roman and Raynal [16], who applied acidified nutrient solutions to sugar maple seedlings. After four days of constant exposure, many of the radicles treated with pH 2.4 and 3.0 solutions darkened and softened at the tip. When these seedlings were planted in greenhouse soil and watered with tap water, only 7 and 53% of the seedlings (at pH 2.4 and 3.0, respectively) survived. However, if exposed seedlings from these treatments were planted in soil and watered with distilled water containing fungicide, all seedlings from all treatments survived.

The occurrence of foliage exposure to acid rain has implications for root diseases as well as foliar diseases. Many conditions that affect aboveground plant parts modify quantities and compositions of root exudates, which in turn cause changes in the biology of the rhizosphere [19]. McGuirt [20] exposed soybean plants three times per week for 40 days to simulated rains of pH 6.0 or 3.5. Assays performed on rhizosphere soils with a range of nutrient agar media indicated that rhizospheres of plants exposed to "rains" of pH 3.5 selected for bacteria that grew quickly on media amended with vitamins, amino acids and other complex nutrients. Apparently, roots of plants exposed to the more acidic treatments exuded more of these complex growth factors. These results are consistent with data reviewed by Moore [21], which indicate that exposure of plant roots to low-pH conditions increases permeability of cell membranes and allows increased leakage of cell contents.

Changes in root exudates resulting from exposure to acid rain would alter microorganism interrelationships involving potential pathogens in the rhizosphere or might alter pathogen responses to the presence of plant roots. Such situations may influence disease occurrence. For example, the charcoal root rot pathogen *Macrophomina phaseolina* (Tassi) Goid. has a host range of more than 400 plant species, and sclerotium populations have been correlated with populations of certain actinomycetes and bacteria in soil [22]. Changes in rhizosphere species composition could decrease or stimulate pathogen antagonists and affect disease incidence accordingly. Marx and Bryan [23] found that sporulation of *Phytophthora cinnamomi* Rands and infection patterns in pine roots depended on qualitative characteristics of rhizosphere bacterial populations.

Chemotactic responses of pathogen spore germ tubes depend on root exudate composition as well [24]. Yung and Hickman [25] have demonstrated that the cationic fraction of pea root exudates caused accumulation of *Pythium aphanidermatum* (Edson) Fitzp. zoospores as well as the crude exudates. Zentmyer [26] demonstrated the relationship among foliage, root exudates and responses by motile spores when he removed plant foliage and found reduced attraction for pathogen zoospores by the roots and fewer infections. Necrosis of foliar tissue due to acid rain or abscission of damaged leaves might produce a similar effect.

Chronic effects of repeated exposures of plants to acid rain could conceivably affect root diseases through nutrient imbalance induced in the host. Nutrient status in plant tissues often plays a role in disease development. Yellowing of pin oak foliage due to iron deficiency is common when the soil pH is >7 and iron becomes unavailable for root uptake. Such nutrient availabilities are among the factors that determine the optimum soil pH ranges for various plant species [27], and plants under nutrient stress are frequently more susceptible to attack by pests of various kinds. Changes in tissue nutrient status might occur due to leaching of minerals from foliage, soil or both, resulting in changes of host–parasite relations. For example, increased calcium concentrations in tomato tissues can lead to increased resistance to vascular wilt caused by the soilborne fungus *Fusarium oxysporum* Schlect. emend. Snyd. and Hans [28]. Mineral deficiencies in *Pinus resinosa* Ait. seedlings apparently contributed to heavy losses due to damping-off by *Fusarium* spp. [29]. Various minerals have been shown to leach from plant tissues and soils exposed to simulated acid rain [30–34]. Effects of acid rain on plant diseases through soil nutrient availability and soil pH changes might become apparent only after years of exposure to long-term crops on mineral soils, i.e., forest trees, due to current rates of acid inputs and buffering capacities of most soils [35].

Parasite-Mediated Effects

Direct effects of acid rain on soilborne plant pathogens could conceivably alter the progress of plant–parasite interactions. Populations of most root pathogens are greatest within 10–15 cm of the soil surface, where they have great impact on feeder roots of established plants and the limited root systems of young seedlings. Indeed, since most seeds germinate within a centimeter or so of the soil surface, pathogens quite near the surface are of extreme importance in the pathology of young plants. It is also in this zone that soil microorganisms are under the greatest influence of rainwater chemistry. Acidic rainwater percolating through the soil profile changes in chemica! characteristics through exchange phenomena in soil, but (depending on soil physical and chemical characteristics) varying soil depths may be required to significantly alter severely acid rain. Near-surface pathogens may thus be influenced in some soils.

The optimum growth and reproduction of a soil microorganism occurs within a defined pH range that is species-specific, which may determine incidence of a given disease. For example, the actinomycete *Streptomyces scabies* (Thaxt.) Waks.

and Henrici is sensitive to acid conditions in vitro, and this probably explains the rare occurrence of potato scab in soils characterized by pH less than 5.4 [36]. Club root of cabbage (caused by *Plasmodiophora brassicae* Wor.), however, is severe in some acid soils and is rare in alkaline soils, probably because the spores of the pathogen germinate poorly when the pH is >7 [37]. Activities of these and other pathogens may be greatly affected when acid water enters the soil. A sudden influx of a pH 3.5 soil solution may halt the infection process by *S. scabies*, for example. Rainwater of this acidity occurs presently in the eastern United States.

Different phases of soilborne plant pathogen life cycles respond differently under different pH conditions. Cameron and Milbrath [38] found that most of the eight *Phytophthora* spp. tested grew best at in vitro pH levels of 4.5–5.5. Chee and Newhook [39] observed that the low pH levels in pH-adjusted soil extracts that would inhibit production of *P. cinnamomi* sporangia probably would also limit plant growth. However, Pegg [40] found fewer sporangia and a lower incidence of pathogen isolation when sulfur was used to adjust infested soil to pH 3.8. A drop in pH from 5.0 to 4.5 caused only an 8% reduction in *P. cinnamomi* chlamydospore germination [41], but the same pH reduction caused a 94% depression of mycelial growth after 28 days [38]. Thus, the phase during which an acid rain event occurs may determine the nature of the ultimate effect on disease.

Some of the relationships between plant roots and motile zoospores of oomycete fungi were briefly mentioned previously. Khew and Zentmyer [42] suggest that tropic responses of zoospores are related to cationic charges in the soil solution surrounding the root. A sudden influx of acidic rainwater might impair normal tropic responses of zoospores to charged substances. If tropic responses of pathogen motile spores are indeed related to cationic charges, the sudden influx of acidic rainwater into the soil might alter normal tactic functions of the pathogen. Allen and Harvey [43] described "negative chemotaxis" of *P. cinnamomi* zoospores exposed to different concentrations of HCl or its salts. Zoospores exhibited abnormal turning movements, movement away from diffusing solutions and encystment. Experiments demonstrated that the response depended on critical concentrations of cations rather than on concentration gradients. For hydronium ions, the critical concentration was estimated at 37 μM.

Effects on various stages of pathogen life cycles could eventually influence their populations in soil. Lawry [44] studied fungal populations in soils of sites exposed to acid runoff from strip mines and found net decreases in fungal species diversities. Shafer et al. [45] filled columns to different depths with fir nursery soil naturally infested with *Phytophthora cinnamomi*. After five consecutive daily exposures to simulated rains (pH 5.6, 4.0, 3.2 or 2.4), average *P. cinnamomi* populations in the columns and the pH of effluent water from the columns varied directly with rain treatment pH and soil depth. Propagules within 4 cm of the soil surface had less chance of survival of exposures to rains of pH 4.0 or less than did propagules at greater depths. In another study, Shafer (unpublished results) found that populations of viable *Macrophomina phaseolina* sclerotia in the top 4 cm of pine nursery soil in pots were 20% lower in soils exposed 25 times over 86 days to simulated rains at pH 2.4 than at pH 3.2, 4.0 or 5.6.

Several reports suggest that many plant-parasitic nematodes frequently occur in slightly acidic soils (pH 5.0–6.0 range). Growth and reproduction by several species seem to decrease as bulk soil pH exceeds 6.0. Growth and reproduction of many species apparently are inhibited beyond the pH 4.0–7.0 range [46]. Data related to nematodes are based on overall soil pH; the effect of mass flow of acidic rainwater in soil on nematodes should receive experimentation. To date, only Shriner [47] has examined the impact of acid rain on a nematode-induced plant disease. In a field study with root-knot of kidney beans, he found that plants exposed to simulated rain of pH 3.2 had only 34% as many *Meloidogyne hapla* Chitwood eggs per root system and 48% fewer root galls, compared to plants exposed to pH 6.0 treatments. Plant size and yield, as well as soil pH, were the same for both groups of plants. Shriner speculated that the free-living larval stages of the nematode may have been affected by short-term acidification of soil water, or that parasitism and reproduction of adult females within the plant roots were inhibited by physiological alterations in the host induced by acid rain treatments.

Effects on Beneficial Root–Microorganism Relationships

Some soilborne microorganisms form beneficial symbioses with plant roots. Although these are not "disease" relationships in the eyes of most plant pathologists, these situations share certain characteristics with their harmful counterparts, such as infection and tissue colonization.

Perhaps the best known of these mutualistic symbioses is the infection and nodulation of legume roots by species of the bacterium genus *Rhizobium*. This relationship allows fixation of atmospheric dinitrogen gas by the root–bacterium system. Effects of acid rain on legume nodules have been the subject of several studies. Shriner [47] found that kidney beans and soybeans exposed over eight weeks in the field or five weeks in the greenhouse to simulated rains of pH 3.2 developed an average of 75% fewer nodules than plants exposed to pH 6 treatments. Overall, 74% of the plants that failed to nodulate at all had received pH 3.2 treatments. Results differed between plants that had either foliage only or soil only exposed, suggesting that at least part of the effect was host-mediated. Plants with fewer nodules exhibited lower nitrogenase activity. In a time-related study, nodulation was inhibited most by daily exposures to pH 3.2 treatments applied 15–21 days after plants were inoculated with *Rhizobium*. Waldron [48] also found inhibited nodulation of soybean and kidney bean plants by simulated rains of pH 3.2, but nitrogenase activity was similar to that of plants exposed to deionized water "rains," suggesting some type of compensatory effect by the plant–bacterium system. Waldron also demonstrated that hydrogen ions rather than sulfate ions were responsible for treatment differences, and that soil drenches were more inhibitory than foliar applications, further suggesting a host-mediated effect.

Mycorrhizae are perhaps the most common plant root–soil microorganism symbiosis in nature. Gerdemann [49] states that, "In nature, the mycorrhizal condition is the rule, the nonmycorrhizal condition, the exception." The various patterns of fungal infections of plant roots characteristic of different mycorrhizal types benefit their host plants by effectively increasing absorptive surface areas of the root system and providing for increased uptake of certain soil minerals (and in some cases, water). Presumably, the fungi involved might respond in ways similar to fungal pathogens to acid water in soil. Feicht [50] planted soybeans into pots containing soil infested with the endomycorrhizal fungus *Glomus macrocarpus* Tul. and Tul. var. *geosporus* (Gerd. and Nichol.) Trappe and Gerd. Plants were exposed in the field to simulated rains over the pH range 2.8–5.5 twice each week for 15 weeks. No changes were observed for *G. macrocarpus* colonization of roots. However, sporulation of the fungus associated with plants exposed to pH 2.8 treatments was inhibited by approximately 40%, compared to plants exposed to pH 5.6 treatments.

Ectomycorrhizae of certain forest tree species have been demonstrated to protect the plants from root diseases by several mechanisms [51]. Effectiveness of several of these, such as provision of a physical barrier to parasite infection and activity of antibiotics, might be reduced by acid rain through physiological effects on the host or by acid water in the soil. For example, damage to the fungal mantle might allow pathogens to breach this protective barrier, or antibiotics produced by the endophyte could be inactivated by low pH conditions.

CONCLUSIONS

Most plant diseases consist of delicate interactions between higher plants and microorganisms. Acidic precipitation represents an environmental stress that has been shown to affect expected development of some diseases and similar phenomena under experimental conditions. From the perspective of the "disease triangle" framework, this impact may be expressed through increased plant susceptibility (through physical wounding or impaired physiological activity), decreased plant susceptibility through altered metabolic pathways, decreased pathogen activity (from intolerance to low-pH conditions by vegetative or survival structures, or from increased activity of competitors) or increased pathogenicity (by stimulation from more favorable pH conditions for acidophilic organisms, or by elimination of competitors). Aerial plant parts are directly exposed to rainfall and the potential influence of acidic precipitation on pathogen and diseases of aboveground tissues seems obvious. However, soilborne pathogens and root diseases may also be significantly altered by short-term acidification of the soil solution resulting from acid deposition, or by gradual changes in bulk soil chemical characteristics over a long period of time. Although research in this area is only beginning, knowledge of the biological, ecological and economic impacts of diseases of agricultural crops and native vegetation suggests that acidic precipitation plays a significant role in the subtle interactions of plants and their parasites.

ACKNOWLEDGMENTS

This work was supported in part by a subcontract from the EPA/NCSU Acid Precipitation Program, EPA Cooperative Agreement CR-806912 between the U.S. Environmental Protection Agency and North Carolina State University. Paper No. 8358 of the journal series of the North Carolina State Agricultural Research Service, Raleigh, North Carolina.

REFERENCES

1. Bateman, D.F. In: *Plant Disease: An Advanced Treatise, Vol. III,* J.G. Horsfall and E.B. Cowling, Eds. (New York: Academic Press, Inc., 1978), pp. 53–62.
2. Campbell, C.L., J.S. Huang and G.A. Payne. In: *Plant Disease: An Advanced Treatise, Vol. V.,* J.G. Horsfall and E.B. Cowling, Eds. (New York: Academic Press, Inc., 1980), pp. 103–118.
3. Purnell, T.J., and R.F. Preece. "Effects of Foliar Washing on Subsequent Infection of Leaves of Swede (*Brassica napus*) by *Erysiphe cruciferarum,*" *Physiol. Plant Pathol.* 1:123 (1971).
4. Martin, J.T., and B.E. Juniper. *The Cuticles of Plants* (New York: St. Martins Press, 1970).
5. Smith, W.H. *Microbiology of Aerial Plant Surfaces* (New York: Academic Press, Inc., 1976).
6. Bruck, R.I., S.R. Shafer and A.S. Heagle. "Effects of Simulated Acid Rain on the Development of Fusiform Rust on Loblolly Pine," *Phytopathology* 71:864 (1981).
7. Smith, W.H. Personal communication.
8. Stapp, C. *Bacterial Plant Pathogens* (London: Oxford University Press, 1961).
9. Shriner, D.S., and E.B. Cowling. In: *Effects of Acid Precipitation on Terrestrial Ecosystems,* T.C. Hutchinson and M. Havas, Eds. (New York: Plenum Publishing Corp. 1980), pp. 435–442.
10. McLaughlin, S.B., and D.S. Shriner. In: *Plant Disease: An Advanced Treatise, Vol. V,* J.G. Horsfall and E.B. Cowling, Eds. (New York: Academic Press, Inc., 1980), pp. 407–431.
11. Umbriet, W.W. *Modern Microbiology* (San Francisco, CA: W.H. Freeman & Company, 1962).
12. Leben, C. "Influence of Acidic Buffer Sprays on Infection of Tomato Leaves by *Alternaria solani,*" *Phytopathology* 44:101–106 (1954).
13. Shriner, D.S. "Effects of Simulated Acid Rain on Host-Parasite Interactions in Plant Diseases," *Phytopathology* 68:213–218 (1978).
14. Lacy, G.H., B.I. Chevone and N.P. Cannon. "Effects of Simulated Acidic Precipitation on *Erwinia herbicola* and *Pseudomonas syringae* Populations," *Phytopathology* 71:888 (1981).
15. Teigen, O. "Spire-og etableringsforsok meal gran og furu i kunotig for suret mineraljord (Experiments with Germination and Establishment of Spruce and Pine in Artificially Acidified Mineral Soil)," SNSF Project IR 10/75 (1975).
16. Roman, J.R., and D.J. Raynal. "Effects of Acid Precipitation on Vegetation," New York State Energy Research and Development Authority, ERDA Report 80–28 (1980), pp. 4–1 to 4–63.

17. Cohen, C.J., L.C. Grothaus and S.C. Perrigan. "Effects of Simulated Sulfuric Acid Rain on Crop Plants," Special Report 619, Agricultural Experiment Station, Oregon State University, Corvallis (1981).
18. Arnon, D.I., and C.M. Johnson. "Influence of Hydrogen Ion Concentration on the Growth of Higher Plants Under Controlled Conditions," *Plant Physiol.* 17:525–539 (1942).
19. Rovira, A.D., and C.B. Davey. In: *The Plant Root and Its Environment,* E.W. Carson, Ed. (Charlottesvile, VA: The University Press of Virginia, 1974), pp. 153–204.
20. McGuirt, P.V., Jr. "Effects of Simulated Rain Acidified with Sulfuric Acid on Forest and Agricultural Ecosystems," MS Thesis, North Carolina State University, Raleigh, NC, (1976).
21. Moore, D.P. In: *The Plant Root and Its Environment,* E.W. Carson, Ed. (Charlottesville, VA: The University Press of Virginia, 1974), pp. 135–151.
22. Dhingra, O.D., and J.B. Sinclair. "Survival of *Macrophomina phaseolina* Sclerotia in Soil: Effect of Soil Moisture, Carbon:Nitrogen Ratios, Carbon Sources, and Nitrogen Concentrations," *Phytopathology* 65:236–240 (1975).
23. Marx, D.H., and W.C. Bryan. In: *Root Diseases and Soilborne Pathogens,* T.A. Toussoun, Robert V. Bega, and Paul E. Nelson, Eds. (Berkeley, CA: University of California Press, 1970), pp. 171–172.
24. Zentmyer, G.A. In: *Root Diseases and Soilborne Pathogens,* T.A. Toussoun, Robert V. Bega, and Paul E. Nelson, Eds. (Berkeley, CA: University of California Press, 1970), pp. 109–111.
25. Yung, C.-H., and C.J. Hickman. In: *Root Diseases and Soilborne Pathogens,* T.A. Toussoun, Robert V. Bega, and Paul E. Nelson, Eds. (Berkeley, CA: University of California Press, 1970), pp. 103–108.
26. Zentmyer, G.A. In: *Ecology of Soil-borne Plant Pathogens,* K.F. Baker and W.C. Snyder, Eds. (Berkeley, CA, University of California Press, 1965), p. 246.
27. Foth, H.D., and L.M. Turk. *Fundamentals of Soil Science* (New York: John Wiley & Sons, Inc., 1972).
28. Jones, J.P. and S.S. Woltz. "Fusarium Wilt (Race 2) of Tomato: Calcium, pH, and Micronutrient Effects on Disease Development," *Plant Dis. Rep.* 53:276–279 (1969).
29. Tint, H. "Studies in the Fusarium Damping-off of Conifers. II. Relation of Age of Host, pH, and Some Nutritional Factors to the Pathogenicity of Fusarium," *Phytopathology* 35:440–457 (1945).
30. Abrahamsen, G., K. Bjor, R. Hontvedt and B. Tveite. In: *Impact of Acid Precipitation on Forest and Freshwater Ecosystems in Norway,* Finn H. Braekke, Ed. (Oslo: SNSF-Project, 1976), pp. 36–63.
31. Fairfax, J.A.W., and N.W. Lepp. "Effect of Simulated 'Acid Rain' on Cation Loss from Leaves," *Nature* 255:324–325 (1975).
32. Nilsson, S. Ingvar. In: *Effects of Acid Precipitation on Terrestrial Ecosystems,* T.C. Hutchinson and M. Havas, Eds. (New York: Plenum Publishing Corp., 1980), pp. 537–551.
33. Wood, T., and F.H. Bormann. "Increases in Foliar Leaching Caused by Acidification of an Artificial Mist," *Ambio* 4:169–171 (1975).
34. Wood, Tim and F.H. Bormann. In: *Proceedings of the First International Symposium on Acid Precipitation and the Forest Ecosystem,* L.S. Dochinger and T.A. Seliga, Eds., USDA Forest Service/General Technical Report NE-23, (Upper Darby, PA: USDA Experimental Station, 1976), pp. 815–825.
35. Frink, C.R., and G.K. Voight. In: *Proceedings of the First International Symposium*

on *Acid Precipitation and the Forest Ecosystem,* L.S. Dochinger and T.A. Seliga, Eds., USDA Forest Service/General Technical Report NE-23, (Upper Darby, PA: USDA Experimental Station 1976), pp. 685–709.

36. Alexander, M. In: *Effects of Acid Precipitation on Terrestrial Ecosystems,* T.C. Hutchinson and M. Havas, Eds. (New York: Plenum Publishing Corp., 1980), pp. 363–374.

37. Agrios, G.N. *Plant Pathology* (New York: Academic Press, Inc., 1978), pp. 196–200.

38. Cameron, H.R. and G.M. Milbrath. "Variability in the Genus *Phytophthora.* I. Effects of Nitrogen Sources and pH on Growth," *Phytopathology* 55:653–657 (1965).

39. Chee, K.H., and F.J. Newhook. "Variability in *Phytophthora cinnamomi* Rands," *New Zealand J. Agric. Res.* 8:96–103 (1965).

40. Pegg, K.G. "Soil Application of Elemental Sulfur as a Control of *Phytophthora cinnamomi* Root and Heart Rot of Pineapple," *Aust. J. Exp. Agric. Anim. Husb.* 17:859–865 (1977).

41. Mircetich, S.M., G.A. Zentmyer and J.B. Kendrick, Jr. "Physiology of Germination of Chlamydospores of *Phytophthora cinnamomi,*" *Phytopathology* 58:666–671 (1968).

42. Khew, K.L., and G.A. Zentmyer. "Chemotactic Responses of Zoospores of Five Species of *Phytophthora,*" *Phytopathology* 63:1511–1517 (1973).

43. Allen, R.N., and J.D. Harvey. "Negative Chemataxis of Zoospores of *Phytophthora cinnamomi,*" *J. Gen. Microbiol.* 84:28–38 (1974).

44. Lawry, J.D. "Soil Fungal Populations and Soil Respiration in Habitats Variously Influenced by Coal Strip-Mining," *Environ. Poll.* 14:195–205 (1977).

45. Shafer, S.R., R.I. Bruck and A.S. Heagle. "Survival of *Phytophthora cinnamomi* in Soil Columns Exposed to Simulated Acid Rain," *Phytopathology* 72:361 (1982).

46. Norton, D.C. "Relationship of Physical and Chemical Factors to Populations of Plant-Parasitic Nematodes," *Ann. Rev. Phytopathol.* 17:279–299 (1979).

47. Shriner, D.S. "Effects of Simulated Rain Acidified with Sulfuric Acid on Host-Parasite Interactions," PhD Thesis, North Carolina State University, Raleigh, NC (1974).

48. Waldron, J.K. "Effects of Soil Drenches and Simulated 'Rain' Applications of Sulfuric Acid and Sodium Sulfate on the Nodulation and Growth of Legumes," MS Thesis, North Carolina State University, Raleigh, NC (1978).

49. Gerdemann, J.W. In: *Mycorrhizae,* USDA—Forest Service Misc. Publication 1189 (1971), pp. 9–18.

50. Feicht, P.G. "Exposure of Soybeans to Ozone or Simulated Acid Rain and Interactions with *Glomus macrocarpus,*" MS Thesis, North Carolina State University, Raleigh, NC (1980).

51. Marx, D.H. "Ectomycorrhizae as Biological Deterrents to Pathogenic Root Infections," *Ann. Rev. Phytopathol.* 10:429–454 (1972).

CHAPTER 4

Effects of Acidic Deposition on Forest Vegetation: Interaction With Insect and Microbial Agents of Stress

William H. Smith
Gordon Geballe
Jurg Fuhrer

This chapter and numerous recent reviews [1–3] have identified a variety of real and potential direct and indirect effects of acidic deposition on terrestrial vegetation. Several important indirect effects to be considered are potential alterations of host–insect interactions, host–parasite interactions and symbiotic associations. Presumably, this relationship could involve a direct influence of acidic deposition on a host plant, a direct influence of acidic deposition on an insect, microbial pathogen or microbial symbiont, or a direct influence of acidic deposition on the interactive process of plant and agent, i.e., infestation, disease or symbiosis.

This chapter will attempt to summarize the literature containing evidence to support or reject the hypotheses that acidic deposition predisposes forest trees to greater insect infestation or greater microbial disease.

INSECTS

Arthropods have roles of enormous importance in the structure and function of terrestrial ecosystems. Forest ecosystems, in particular, typically have large and diverse arthropod populations. The damaging influence of high population densities of certain insects can be very visible and cause widespread forest destruction; witness contemporary North American situations involving the Douglas fir tussock moth, the gypsy moth, the eastern spruce budworm and the southern pine bark beetle. It is critically important, however, to maintain a perspective that there is substantial evidence to support the hypothesis that forest insects, even those that cause massive destruction in the short run, may play essential and beneficial roles in forest ecosystems in a long-term context. These roles may involve regulation of tree species competition, species composition and succession, primary production, and nutrient cycling [4,5]. As a result, an assessment of the interrelationships between acidic precipitation and phytophagous insects is important.

There is increasing indication that a variety of particularly damaging forest insects detect and respond to stress-induced alterations in host tree physiology. The stresses are variable and may include microbial infection, climatic extremes, edaphic factors, fire and age. Massive insect infestations are characteristically initiated in middle-aged to mature forests typified by reduced productivity rates. Localized and scattered insect outbreaks are associated with forests of all ages, but are generally associated with the least vigorous trees with slow growth rates. Some investigators judge that certain insect population growth is inversely related to host plant vigor [5].

It is essential, therefore, to appreciate the interactions between air pollutants and forest insects because of the critical importance of these animals to forest ecosystem structure and function. Air contamination in general, and acidic deposition in particular, may be an additional environmental stress factor capable of predisposing forest trees to detrimental arthropod influence. Research dealing with insect–air pollution interactions, however, is not extensive, despite the fact that insect–air pollution research has a history that extends over 50 years. European literature dealing with this topic is substantially larger than North American literature. Both data bases are oriented toward an understanding of gaseous pollutant effects, and little is known about the impact of acidic deposition.

A variety of studies have presented data indicating that species composition or population densities of insect groups are altered in areas of high air pollution stress, e.g., roadside [6] or industrial [7–9] environments. Specific information is available on the general influence of polluted atmospheres on population characteristics of forest insects [10–21]. Johnson [22,23] has reviewed much of the literature dealing with air pollutants and insect pests of conifers. One of the most comprehensive literature reviews available concerning forest insects and air contaminants has been presented by Villemant [24]. Alstad et al. [25] have provided an excellent overview of the effects of air pollutants on insect populations.

PATHOGENS

Abnormal physiology, or disease, in woody plants follows infection and subsequent development internally or on the surface of tree parts by an extremely large number and diverse group of microorganisms. All stages of tree life cycles and all tree tissues and organs are subject, under appropriate environmental conditions, to impact by a heterogeneous group of microbial pathogens including viroids, viruses, mycoplasmas, bacteria, fungi and nematodes. The influence of a specific disease on the health of an individual tree may range from innocuous to mild to severe.

Over extended time periods, the interaction of native pathogens with natural forest ecosystems is significant and frequently beneficial to ecosystem development and function. As in the instance of insect interactions, microbes and the diseases they cause play important roles in succession, species composition, density, competition and productivity. In the short term, the effects of microbial pathogens may conflict with forest management objectives and assume a considerable economic or managerial as well as ecological significance [26].

The interaction between air pollutants and microorganisms in general is highly variable and complex. Babich and Stotzky [27] and Laurence [28] have provided comprehensive overviews of the subject. Microbes may serve as a source as well as a sink for air pollutants. A specific air pollutant, at a given dose, may be stimulatory, neutral or inimical to the growth and development of a particular virus, bacterium or fungus. In fungi, fruiting body formation, spore production and spore germination may be stimulated or inhibited. Microorganisms that normally develop in plant surface habitats may be especially subject to air pollutant influence by virtue of their exposure. These microbes have received considerable research attention and have been the subject of review [29–32]. The apparent influence of individual pollutants and combinations of pollutants on microbial plant parasites is to both increase and decrease their activities. The actual impact of air pollution stress on disease expression is especially complicated, since air contaminants not only influence the metabolism and ecology of the microbe but also influence the physiology of the host plant. Even under "unpolluted atmospheric conditions," disease in plants is a complex integration of pathogen physiology, host plant physiology and ambient environmental conditions. The addition of an air pollutant stress has the effect of adding an additional variable to an already elaborate and complicated interaction. Numerous comprehensive reviews have summarized the interactions between air contaminants and plant diseases. Heagle [33] summarized nearly 100 references and found that sulfur dioxide, ozone or fluoride had been reported to increase the incidence of 21 diseases and decrease the occurrence of nine diseases in a variety of nonwoody and woody hosts. Treshow [34] has provided a detailed review concerning the influence of sulfur dioxide, ozone, fluoride and particulates on a variety of plant pathogens and the diseases they cause. Treshow lamented the fact that most of the data available deal with in vitro or laboratory accounts of microbe–air pollutant interactions, while only a few investigations have examined the influence of air pollutants on disease development under field conditions. In a review provided by Manning [35], it was pointed out that most research attention has been directed to fungal pathogen–air pollutant interactions. Greater research perspective is needed concerning air pollution influence on viruses, bacteria, nematodes and the diseases they cause. Macroscopic agents of disease, most importantly true- and dwarf-mistletoes, must also be examined relative to air pollution impact, especially in the western part of North America where the latter are extremely important agents of coniferous disease.

Forest trees, because of their large size, extended lifetimes and widespread geographic distribution, are subject to multiple microbially-induced diseases frequently acting concurrently or sequentially. The reviews of Heagle [33], Treshow

[34] and Manning [35] included consideration of a variety of pollutant–woody plant pathogen interactions, but were not specifically concerned with forest tree disease. In their review of the impact of air pollutants on fungal pathogens of forest trees of Poland, Grzywacz and Wazny [36] referenced literature citations indicating that air pollution stimulated the activities of at least 12 fungal tree pathogens while restricting the activities of at least 10 others.

Our understanding of the influence of acidic deposition on pathogens and the diseases they cause is very meager. Shriner [37–39] has provided us with some very valuable perspectives in this important but understudied area. Falling precipitation and the precipitation wetting of vegetative surfaces play an enormously important role in the life cycles of a large number of plant pathogens. In recognition of this, Shriner examined the effects of simulated rain acidified with sulfuric acid on several host–parasite systems under greenhouse and field conditions [37–39]. The simulated precipitation he employed had a pH of 3.2 and 6.0, approximating the common range of ambient precipitation pH.

The application of simulated precipitation of pH 3.2 resulted in (1) an 86% restriction of telia production by *Cronartium fusiforme* (fungus) on willow oak; (2) a 66% inhibition of *Meloidogyne hapla* (root-knot nematode) on kidney bean; (3) a 29% decrease in percentage of leaf area of kidney bean affected by *Uromyces phaseoli* (fungus); and (4) both stimulated and inhibited development of halo blight of kidney bean caused by *Pseudomonas phaseolicola* (bacterium). In the latter case, the influence of acidic precipitation varied and depended on the particular stage of the disease cycle when the exposure to acidic precipitation occurred; simulated sulfuric acid rain applied to plants before inoculation stimulated the halo blight disease by 42%. Suspension of inoculum in acidic precipitation decreased inoculum potential by 100%, while acidic precipitation applied to plants after infection occurred inhibited disease development by 22% [37].

Examination of the willow oak and bean leaves using scanning electron microscopy revealed distinct erosion of the leaf surface by rain of pH 3.2. This may suggest that altered disease incidence may be due to some change in the structure or function of the cuticle. Shriner has also proposed that the low-pH rain may have increased the physiological age of exposed leaves. Shriner [40] concluded his initial experiments by suggesting that he had not established threshold pH levels at which significant biological ramifications to pathogens occur from acidic precipitation. He did suggest, however, that artificial precipitation of extremely low pH probably alters infection and disease development of a variety of microbial pathogens.

INFLUENCES ON PLANT HOSTS THAT WOULD ALTER INITIAL HOST RELATIONSHIPS WITH INSECTS OR PARASITES

After conditions of sufficient dose, air pollutants directly cause visible injury to forest trees. Droplets of rain with very low pH may cause necrotic spotting of tree leaves. Simulated acidic rain has been shown to produce visible foliar injury of angiosperm and gymnosperm plants if the pH is less than approximately 3.5 (316 μeq-L^{-1} H$^+$) [41]. The occurrence of this event in natural environments is judged to be rare. In addition, responses of forest vegetation to acidic precipitation are highly variable. The most important reasons include differences in genetic constitution and age of trees along with variation caused by many environmental conditions. Under experimental conditions, direct effects on plants by acidic precipitation that have been reasonably well documented include [42–45]:

1.   damage to protective surface structures such as the leaf cuticle;
2.   leaching of organic and inorganic compounds from foliage;
3.   interference with metabolic processes;
4.   interference with plant growth and yield; and
5.   influence on reproductive processes.

Exposure to acidic precipitation may lead to acidification of plant surfaces, especially leaves and bark [46]. Leaf cuticles may be eroded by chronic exposure to acid precipitation [47]. Evans and Curry [48] demonstrated that initial injury from acidic rain preferentially affects the leaf indumentum near trichomes and vascular tissues. Repeated daily exposures of plants to simulated acidic rain can result in hypertrophy of mesophyll cells in *Tradescantia* sp. [48] and *Populus* sp. [49]. Lesions were most frequent on lateral edges of leaves and at vascular bundles [50]. These effects are likely to influence phytophagous insects and microbial pathogens associated with plant surfaces. Foliar lesions could release plant volatiles attractive or repulsive to insect pests. The lesions could also serve as infection courts for microbial disease agents. Lesion location near vascular bundles may be especially efficient for increasing infection by vascular pathogens.

Materials washed or leached from foliage by rain, mist and dew include a great variety of compounds. Inorganic materials leached include all the macro- and microelements essential for plant growth along with nonessential elements. Organic materials leached include sugars, pectic substances, sugar alcohols, amino acids, growth-regulating compounds (gibberellins, auxins, abscisic acid), vitamins, alkaloids and phenolic substances [51]. The influence of leached chemicals on insects infesting tree leaves or bark could be attractive, repulsive or provide chemical orientation. In the case of surface microbes, leached compounds may inhibit vegetative growth or spore germination (alkaloids, phenolic substances) or stimulate vegetative growth (as nutrients) or spore germination (as inducers or nutrients—sugars, amino acids, vitamins). Leaching of toxic radioelements from plant surfaces could have a restrictive impact on plant surface biota [52].

Growth and yield of plants may be stimulated or inhibited by acidic precipitation. If growth is either stimulated or suppressed, it is probable that differential influence on insects and pathogens would follow. In the case of some host–pathogen and host–insect relationships a tree under stress is more vulnerable to infestation or infection. Bark beetles, and root-infecting and canker-forming fungi are generally more successful in less vigorous individuals. Trees exhibiting vigorous growth, on the other hand, may be predisposed to more serious impact from certain rust fungi and other disease agents.

INFLUENCE ON INSECTS

Air pollutants may directly affect insects by influencing growth rates, mutation rates, dispersal, fecundity, mate finding, host finding and mortality. Indirect effects may occur through changes in host age structure, distribution, vigor and acceptance. Few researchers have investigated the effects of acidic deposition on insects. Some studies relative to acidity effects on aquatic insects are available [53]. Terrestrial arthropods, on the other hand, have been the subject of very few studies. Hagvar et al. [54] concluded that acidic precipitation, probably from western and central Europe, increases the susceptibility of Scotch pine forests to the pine bud moth (*Exoteleia dodecella*). The evidence available to support this conclusion was modest.

Two host-insect situations might be especially vulnerable to impact by acidic deposition: (1) foliar and other surface-feeding phytophagous insects, and (2) litter- and soil-inhabiting microarthropods. Additional associations of high potential vulnerability are those involving predaceous and parasitoid insects and arachnids (including phytophages) via direct deposit of pollutants on cuticle and eggs, and internal feeders in wood and in galls and leaf mines, if there are significant alterations induced in host metabolism or in secondary plant chemicals. Very little is known at present of the consequences of acidic precipitation for leaf-feeding insects. Due to great differences in mouthpart styles, studies will need to include significant tests of the difference between phytophages with chewing mouthparts (primarily larvae of lepidoptera, sawflies and chrysomelid beetles, all stages of Arcrididae and Tettigoniidae, and adults of several beetle families, notably Scarabeidae, Meloidae and Chrysomelidae) and those with piercing and sucking mouthparts that allow them to by-pass compounds on plant surfaces (primarily Hemiptera, Thysanoptera and phytophagous mites). The former all ingest entire leaves and sometimes other plant structures; therefore, any deposits on plant surfaces enter the alimentary tract. The latter ingest only the internal fluids, and a reasonable working hypothesis is that aphids, scales, mirid bugs, thrips and tetranychid and eriophyid mites are relatively unaffected by leaf-surface pollutants.

A meaningful experimental study must compare the effects on a few well-chosen arthropods with: (1) different phyletic ancestry; (2) different feeding mechanisms; and (3) different body size (both early vs late instar effects and tiny vs large species). In addition, several developmental stages must be examined. First are the rapid effects from feeding on treated foliage by observations during the first few hours of feeding. These may include feeding inhibition such that polluted foliage is behaviorally rejected. Second are the longer-term lethal effects from accumulation of any toxicity. Third are effects on orderly ontogeny; mainly slowed development in general or in particular instars. Fourth are effects on reproductive physiology (fertility and fecundity) and behavioral performance of adults. Each of these is phrased in terms of deleterious effects, but of course it is possible that pH and certain ions will, in fact, improve environmental and food conditions for some arthropods.

It is essential to determine if acidic precipitation can induce alterations in leaf chemistry under field conditions. Recent evidence suggests drought stress increases foliar polyphenol levels in trees. Ethylene, produced by injured leaves, may increase polyphenol levels in adjacent leaves. Ethylene production modification as a result of acidic deposition has never been examined. Leaf tannin increases apparently make oak leaves less palatable to gypsy moth larvae. Defoliation appears to elevate tannin levels in succeeding growing seasons.

INFLUENCE ON MICROBIAL PATHOGENS

As in the instance of insects, microorganisms that cause foliar disease or that infect plants through the leaves may be expected to be especially subject to influence by acidic deposition. This is true for at least three reasons:

1. These microbes may grow and develop vegetatively and reproductively in environments with relatively high levels of ambient pollutant concentrations (for example, leaf surface).
2. Foliar tissue is known to be the site of primary accumulation of recalcitrant materials from the atmosphere, for example, flouride and heavy metals.
3. Foliar tissue is the primary site of direct acidic deposition influence on the plant and leaf tissue may be expected to be predisposed to infection, assuming that physical or metabolic resistance mechanisms are adversely influenced.

As in the case of insects and acidic precipitation, the relationship between acidic rain and pathogens is incompletely understood.

The experiments of Shriner emphasize that acidic precipitation must be evaluated with regard to its ability to influence agents of plant disease and to influence the disease process itself. Substrate pH has long been appreciated to be a critically important determinant of microbial growth and development. In the case of fungal disease agents of forest trees, ambient pH has been shown to be important for growth and sporulation [40], infectivity [55] and microbial antagonism [56].

In recent years, a very serious disease of hard pines caused by a twig and leaf pathogen *Gremmeniella abietina* has increased in importance in northeastern United States. The disease, termed Scleroderris canker, was first reported on red pine in New York in 1964. Currently *G. abietina* is causing significant large tree mortality in Vermont and New York. Since it may be more than coincidence that this region is included within the highest acidic precipitation zone of North America, Raynal et al. [44] recently initiated an acidic rain/Scleroderris research project. The laboratory and field studies reported to date indicate the disease may be affected by precipitation pH, but there was no indication that acidified rain increased disease incidence. In fact, the opposite may be true. That is, acidic rain may reduce the importance of the canker disease [44].

Armillaria mellea is an extremely important forest tree root pathogen throughout the temperate zone. The fungus is geographically very widespread, has an extremely broad host range and is especially significant in causing disease of trees under stress. Shields and Hobbs [55] have indicated that soil pH is related to disease development caused by *Armillaria mellea*. If acidic deposition influences soil pH or tree vigor, it may indirectly change tree susceptibility to *Armillaria mellea* infection. In the northeast, spruce decline in high-elevation forests has been a recent concern. *Armillaria mellea* is associated with spruce exhibiting dieback and decline symptoms in northern New England and may be importantly involved in the morbidity and mortality of this species. The habitats of soil pathogens such as *Armillaria mellea* are buffered to a greater degree than are those of plant-surface habitats; for acid deposition to influence soil pathogens an alteration of soil pH or chemistry or host susceptibility would have to occur.

Fusiform rust caused by *Cronartium fusiforme* causes the most important disease of managed pine in the southeast. Bruck et al. [57] applied simulated rain of various pH levels to loblolly pine at the time of inoculation with rust basidiospores. Significantly fewer galls formed on trees treated with simulated rain at pH 4.0 or less than on trees treated with rain at pH 5.6.

A variety of bacterial species are important components of tree leaf microflora. Lacy et al. [58] observed that populations of *Erwinia herbicola* and *Pseudomonas syringae* were reduced on soybean leaves when host plants were treated with water acidified to pH 3.4 relative to leaves exposed to distilled water (pH 5.7).

LITERATURE REVIEW SUMMARY

At present levels of acid deposition, we can reasonably conclude that there is no direct, acute pathological stress on forest trees. We cannot, however, rule out the possibility of indirect, subtle interaction of acid deposition with phytophagous insects and microbial pathogens. This interaction could result in a stimulation *or* a restriction of pest activity. The former has higher risk potential and must be examined more thoroughly.

The literature supports the following potential mechanisms for acidic deposition impact on predisposition of forest trees to increased disease caused by microbial pathogens and increased infestation caused by phytophagous insects (Figure 1).

1. Alteration of tree morphology, physiology, metabolism or chemistry by acidic deposition may predispose forest trees to enhanced pathogen or insect activity.
2. Acidic deposition may directly stimulate or benefit phytophagous insects such that insect stress on host trees is increased.
3. Acidic deposition may directly stimulate or benefit microbial pathogens such that disease stress on host trees is increased.

Each of these mechanisms may operate over short-term (event, hours, days) and long-term (months, years) time periods. Mechanism 1 is probably more likely to

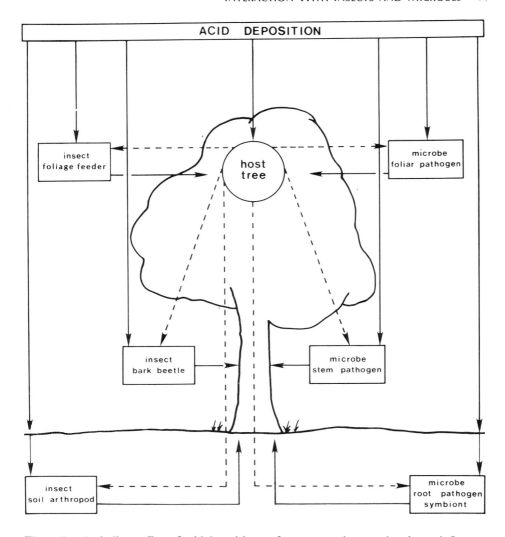

Figure 1.   An indirect effect of acid deposition on forest vegetation may involve an influence on phytophagous insects and disease processes. This relationship could involve a direct influence of acid deposition on host plants, e.g., predisposition (dashed lines), a direct influence of acid deposition on insects or microbial pathogens (solid lines), or a direct influence of acid deposition on the interactive process of plant and stress agent (e.g., infestation or disease). With leaf- or bark-infesting or -feeding insect species, or foliar or stem pathogens, the potential for interaction with acid deposition is particularly high, as these pest organisms spend important segments of their life cycles on the surfaces of plants. Soil inhabiting arthropods and root infecting pathogens are less likely to be influenced by acid deposition than are surface organisms. Soil organism influence would only follow acid deposition induced alteration in soil pH or chemistry *or* altered host resistance.

result from long-term interaction of acidic precipitation with forest systems. Mechanisms 2 and 3 may be more common in response to short-term interaction.

Predisposition of tree hosts may be caused by erosion or other damage to the cuticle. Integrity of cuticular materials and epidermal cells is critically important as a resistance mechanism for a variety of insect and microbial pests. Alteration of foliar volatile release may have important implications for insect host finding. Changes in leaf chemistry, especially those involving ethylene, phenolic compounds and low-molecular-weight sugars and amino acids could have important implications for insect feeding, reproduction and microbial infection. Alterations in foliar leachates caused by acidic precipitation could have important consequences for insect and microbial nutrition and biotic competition and antagonism.

Direct influence of acidic deposition on insect pests, especially those feeding on foliage, bark or flower parts, could include changes in stage-specific morbidity and mortality, fecundity, sex ratios, feeding habits and population dynamics. Direct influence of acidic deposition on microbial pathogens, again especially those infecting aboveground tree surfaces, could include changes in vegetative growth, spore production, spore germination, inoculum potential, infection and epidemiology.

RESEARCH IN PROGRESS

It is our judgment that pathogens that infect stem or foliar tissue are most likely to be influenced by acid deposition. Microorganisms that induce branch and stem cankers and foliar disease, e.g., anthracnose pathogens, have significant portions of their life cycles on plant surfaces and are directly exposed to incident precipitation and stem and leaf runoff. Anthracnose diseases of deciduous tree species are widespread throughout the eastern United States. These diseases are caused by several species of closely related fungi. The prevalence and severity of disease are governed by weather conditions—frequent rains and cool temperatures favor rapid spread. Symptoms include defoliation and damage to young twigs, buds and fruits. Repeated loss of foliage reduces growth, weakens the tree and may increase susceptibility to infection by other microbes, insect attack, winter injury or soil stresses. Species of eastern hardwoods infected by anthracnose fungi include sycamore, London plane tree, oak, walnut, maple, ash, hickory, elm, birch, catalpa, basswood, tulip tree and horse chestnut [59,60]. During the last decade, increases in the significance of anthracnose diseases have been reported in northeastern states [61,62]. The optimum hydrogen-ion concentration for germination of anthracnose spores has been reported to be pH 4–5 in citric acid–phosphate buffer [63].

In order to test the hypothesis that acid precipitation facilitates anthracnose disease development by stimulating the pathogen or predisposing the host, we have constructed an artificial rain device (Figure 2) to apply rain simulants to host plants. We are attempting to mimic, as closely as practical, natural rain. This is difficult as raindrops vary in size, pH varies from storm to storm and within a single storm event, and rainfall occurs at irregular intervals and falls at varying rates. Our rain chamber is an adaption of the basic design developed by

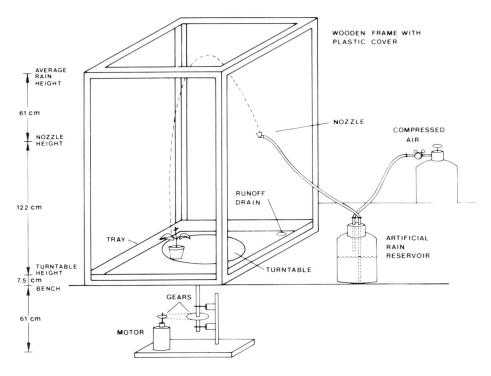

Figure 2. Schematic of rain chamber employed to apply rain simulants to experimental plants. This device effectively reproduces droplet size and terminal velocity of "light rain" conditions. It allows several plants to be uniformly treated concurrently.

Evans and co-workers of the Brookhaven National Laboratory. The chamber effectively simulates "light rain" conditions [64].

Rain simulants were prepared using the cation and anion composition presented by Likens et al. [65] for central New Hampshire. The ionic concentrations were averaged for April through October to be representative of conditions over the vegetative growing season. Simulants of various pH were obtained by adding appropriate amounts of sulfuric, nitric and hydrochloric acid. The chemical composition of rain simulants is shown in Table 1.

Rain simulants of pH 3.0, 4.0 and 5.6 were applied to experimental plants. Plants received three treatments 24 h apart; each rain event was 6 min in duration. Rotation of plants in the rain chamber ensured that all plants received the same amount of rain (2 mm per unit area). If plants were growing in soil, plastic covers were fitted over the pots to exclude rain simulants from the soil. If plants were growing in vermiculite, the pots were flooded with water immediately after the rain treatment to dilute any simulant that entered the pot.

Our initial efforts have concentrated on three host–pathogen systems: anthracnose disease of bean (*Phaseolus vulgaris*) caused by *Colletotrichum lindemuthianum;*

Table I. Chemical Composition (μeq-L^{-1}) of Rain Simulants

	pH			
Ion	5.6[a]	4.0[b]	3.0[b]	4.06[c]
NH_4^+	14.5	14.5	14.5	14.5
Ca^{2+}	9.5	9.5	9.5	9.5
Mg^{2+}	4.1	4.1	4.1	4.1
Na^+	5.2	5.2	5.2	5.2
K^+	1.8	1.8	1.8	1.8
SO_4^{2-}	13.6	97.8	855.6	77.8
NO_3^-	14.5	32.0	189.5	25.4
Cl^-	7.0	17.5	117.0	18.5
H^+	0.0	112.7	1127.0	87.6
Cations	35.1	147.8	1162.1	122.1
Anions	35.1	147.7	1162.1	116.7

[a] Ionic composition of pH 5.6 simulant is the average of the April–October means at Hubbard Brook Experimental Forest in central New Hampshire [65], minus H^+ and a corresponding amount of SO_4^{2-}, NO_3^- and Cl^-.
[b] pH 4.0 and 3.0 simulants were adjusted by adding aliquots of stock acids.
[c] Monthly mean pH for April–October, Hubbard Brook [65]. Difference between cations and anions balanced by organic compounds.

anthracnose disease of sycamore (*Platanus occidentalis*) caused by *Gnomonia platani* and anthracnose disease of white oak (*Quercus alba*) caused by *Gnomonia quercina*. The bean system was employed primarily to allow the development of experimental protocols using easily handled materials. Bean plants of the Black Valentine variety were employed along with the virulent (β) and avirulent (γ) strains of *C. lindemuthianum*.

Work completed to date has included treatment of bean, sycamore and oak seedlings with rain simulants of pH 3.0, 4.0 and 5.6. Treated leaves were examined for visible symptoms, autofluorescence, chlorophyll concentration and peroxidase activity. Bean plants were treated with rain simulant at age 14, 15, and 16 days and were inoculated on day 16 with spores of both the virulent and avirulent strains of *C. lindemuthianum*.

Results

The only visible symptoms observed following treatment of sycamore and white oak seedlings and bean plants with rain simulants of pH 3.0, 4.0 and 5.6 were on bean at the pH 3.0 treatment (Table II). The symptoms on bean consisted of foliar spots of variable size and of silver or tan coloration. The spots appeared within 24 h of the initial rain treatment. Subsequent rain events increased the size and quantity of spots. Three sequential rain events could produce holes in leaf lamina.

Table II.   Influence of Rain Simulants on the Expression of Visible Foliar Symptoms[a] on Treated Plants[b]

Species	No Rain	pH		
		5.6	4.0	3.0
Sycamore				Trace
White Oak				Trace
Bean			Trace	Considerable

[a] Symptoms consisted of necrotic or discolored spots on the leaf.
[b] All plants received three treatments of 6-min duration, 24 h apart. Foliar symptoms were observed 24 h after the final rain event.

Bean leaf discs cleared in ethanol were observed with the fluorescence microscope (excitation wavelength was 360 nm). Tissue exposed to pH 3.0 and 4.0 rain simulants had small (6-cell-diameter) groups of cells with autofluorescent walls. These zones were not seen on untreated or pH 5.6-treated plants. Bean leaf spots, selected at random on pH 3.0-treated plants, were similarly autofluorescent. These autofluorescent zones may represent lignification of the walls of cells damaged by acidity in the rain simulants. These results suggest that pH 4.0 rain simulant does affect the leaf surface in the absence of *visible* symptoms.

Total peroxidase activity was also observed in treated bean leaves. Peroxidase often changes both qualitatively and quantitatively following wounding or infection [66]. At certain doses, ozone, sulfur dioxide and nitrogen dioxide appear to stimulate peroxidase activity [67,68]. Our results (Table III) do not allow any conclusion regarding peroxidase response until the experiments are repeated.

Our most interesting results were obtained from the bean inoculation study. Bean plants treated with pH 3.0 rain simulant and inoculated with the virulent strain of the pathogen exhibited fungal disease symptoms sooner than did plants treated with pH 4.0 simulant. Even more interesting was the observation that

Table III.   Influence (Change in Absorbance Units at 470 nm-g fresh w^{-1}-min^{-1}) of Rain Simulants on the Peroxidase Activity of Three Bean Varieties[a]

Variety	No Rain	pH		
		5.6	4.0	3.0
Black Valentine	69	54	60	42
Burpee's Stringless	69	45	48	18
Taylor's Horticultural	212	210	180	162

[a] Results are from frozen phosphate fractions taken when the plants were 20 days old, 4 days after the last rain treatment.

plants treated with pH 3.0 rain simulant and inoculated with the *avirulent* strain also became diseased. The lowest pH treatment appeared to alter host resistance.

Our results to date are preliminary and incomplete. Our ultimate goal is to examine numerous host–pathogen and host–insect interactions with acid deposition under both controlled and natural (field) environmental conditions.

Our research is concentrating primarily on the anthracnose foliar diseases and the diffuse and perennial canker diseases. Specific host–pathogen systems that will receive investigation include bean anthracnose caused by *C. lindemuthianum,* sycamore anthracnose caused by *Gnomonia platani* and oak anthracnose caused by *Gnomonia quercina.* Canker diseases that will receive study include: hypoxylon canker of aspen caused by *Hypoxylon pruinatum,* chestnut blight of American chestnut caused by *Endothia parasitica,* and white ash twig and branch canker caused by *Cytophoma pruinosa* and *Fusicoccum* sp.

CONCLUSIONS

A review of the evidence on the interaction of forest trees, insect and microbial pests, and acid deposition does not allow generalized statements concerning stimulation or restriction of biotic stress agents or their activities by acid deposition. The modest evidence available is from laboratory and controlled environment studies. There is no evidence on this topic from studies employing large trees under field conditions.

We have presented several testable hypotheses for potential mechanisms of interaction between host plants, biotic stress factors and acid deposition. In the instance of leaf- or bark-infesting or -feeding insect species, or foliar or stem pathogens, the potential for significant interaction with acid deposition is particularly high as these pest organisms spend important segments of their life cycles on the surfaces of plants. Phytophagous soil arthropods and root-infecting microorganisms are less likely to be influenced by acid deposition than are surface organisms. Soil organism influence would only follow acid deposition induced alteration in soil pH or chemistry or altered host resistance.

The potential for biotic stress agent interaction is judged to be higher in forest ecosystems than in agricultural ecosystems not receiving intensive pesticide applications, as pest management would overwhelm and generally preclude meaningful acid deposition–biotic stress interaction.

The threshold range of ambient pH above which little adverse influence on insect or microbial pest activity is anticipated is pH 3.5.–4.0.

ACKNOWLEDGMENT

The research reported here was supported by funds provided by the USDA Forest Service, Northeastern Forest Experiment Station.

REFERENCES

1. Glass, N.R. "Environmental Effects of Increased Coal Utilization: Ecological Effects of Gaseous Emissions from Coal Combustion," *Environ. Health Persp.* 33:249–272 (1979).

2. Cowling, E.B., and L.S. Dochinger. "Effects of Acidic Precipitation on Health and Productivity of Forests," in *Effects of Air Pollutants on Mediterranean and Temperate Forest Ecosystems,* R.R. Miller, Ed., USDA Forest Service, Gen. Tech. Rep. No. PSW-43, Berkeley, CA (1980), pp. 165–173.

3. Smith, W.H. *Air Pollution and Forests* New York: Springer-Verlag, (1981).

4. Huffaker, C.B. "Some Implications of Plant-Arthropod and Higher-Level, Arthropod-Arthropod Food Links," *Environ. Entomol.* 3:1–9 (1974).

5. Mattson, W.J., and N.D. Addy. "Phytophagous Insects as Regulators of Forest Primary Production," *Science* 190:515–522 (1975).

6. Przybylski, Z. "The Effects of Automobile Exhaust Gases on the Arthropods of Cultivated Plants, Meadows and Orchards," *Environ. Poll.* 19:157–161 (1979).

7. Sierpinski, Z. "Influence of Industrial Air Pollutants on the Population Dynamics of Some Primary Pine Pests," *Proc. 14th Cong. Int. Union For. Res. Organiz.* 5(24):518–531 (1967).

8. Novaskova, E. "Influence des Pollutions Industrielles sur les Communautés Animals et L'utilisation des Animaux Comme Bio-indicateurs," in *Influence of Air Pollution on Plants and Animals* (1969), pp. 41–48.

9. Lebrun, P. "Effects Écologiques de la Pollution Atmosphérique sur les Populations et Communautés de Microarthropodes Corticoles (Acariens, Collemboles, et Ptérygotes)," *Bull. Soc. Ecol.* 7:417–430 (1976).

10. Templin, E. "On the Population Dynamics of Several Pine Pests in Smoke-Damaged Forest Stands," *Wissenschaft. Z. Tech. Univ. Dresden* 113:631–637 (1962).

11. Schnaider, Z., and Z. Sierpinski. "Dangerous Condition for Some Forest Tree Species from Insects in the Industrial Region of Silesia," Prace Instytut Badawczy Tesnictwa (Warsaw) No. 316, (1967), pp. 113–150.

12. Sierpinski, Z. "Economic Significance of Noxious Insects in Pine Stands Under the Chronic Impact of the Industrial Air Pollution," *Sylwan* 114:59–71 (1970).

13. Sierpinski, Z. "Secondary Noxious Insects of Pine in Stands Growing in Areas with Industrial Air Pollution Containing Nitrogen Compounds," *Sylwan* 115:11–18 (1971).

14. Sierpinski, Z. "The Economic Importance of Secondary Noxious Insects of Pine on Territories with Chronic Influence of Industrial Air Pollution," *Mitt. Forstl. Bundesversuchsanst Wien* 97:609–615 (1972).

15. Sierpinski, Z. "The Occurrence of the Spruce Spider (*Paratetranychus (Oligonychus) ununquis* Jacoby) on Scotch Pine in the Range of the Influence of Industrial Air Pollution," Institute Badawczego Lesnictwa, Warsaw, Bull. No. 433–434 (1972), pp. 101–110.

16. Boullard, B. "Interactions Entre les Pollutants Atmosphériques et Certains Parasites des Essences Forestiéres. (Champignons et insects)," *For. Privée,* 94:31,33,35,36 (1973).

17. Wiackowski, S.K. and L.S. Dochinger. "Interactions Between Air Pollution and Insect Pests in Poland," 2nd Int. Cong. Plant Pathol., Univ. of Minnesota, Minneapolis, MN, Abstr. No. 0736 (1973), p. 1.

18. Hay, C.J. "Arthropod stress," in *Air Pollution and Metropolitan Woody Vegetation,* W.H. Smith and L.S. Dochinger, Eds., USDA Forest Service, Publica No. PIEF-PA-1, Upper Darby, PA (1975), pp. 33–34.

19. Charles, P.J., and C. Villemant. "Modifications des Nivereaux de Population D'insectes dans les Jeunes Plantations de Pins Sylvestres de la Forët de Roumare (Seine-Maritime) Soumises à la Pollution Atmosphérique," *C.R. Acad. Agric. Fr.* 63:502–510 (1977).

20. Sierpinski, Z., and J. Chlodny. "Entomofauna of Forest Plantations in the Zone of Disastrous Industrial Pollution," in *Relationship Between Increase in Air Pollution Toxicity and Elevation Above Ground,* J. Woldk, Ed. (Warsaw: Institute Badawczego Lesnictwa, 1977), pp. 81–150.

21. Dahlsten, D.L., and D.L. Rowney. "Influence of Air Pollution on Population Dynamics of Forest Insects and on Tree Mortality," in *Effects of Air Pollutants on Mediterranean and Temperate Forest Ecosystems,* P.R. Miller, (Ed.), USDA Forest Service, Gen. Tech. Rep. No. PSW-43, Berkeley, CA (1980), pp. 125–130.

22. Johnson, P.C. "Entomological Aspects of the Ponderosa Pine Blight Study, Spokane, Washington," USDA Bur. Entomol. and Plant Quar., Forest Insect Laboratory, Coeur d'Alene. ID (1950).

23. Johnson, P.C. "Atmospheric Pollution and Coniferophagous Invertebrates," Proc. 20th Ann. Western For. Insect Work Conf., Coeur d'Alene, ID (1969).

24. Villemont, C. "Modifications de L'enclomocemose due Pin Sylvestre en Liaison avec la Pollution Atmospherique en Fôret de Roumare (Seine-Maritime)," Doctoral Dissertation, Pierre and Marie Curie University, Paris, (1979).

25. Alstad, D.N., G.F. Edmunds, Jr. and L.H. Weinstein. "Effects of Air Pollutants on Insect Populations," *Ann. Rev. Entomol.* 27:369–384 (1982).

26. Smith, W.H. *Tree Pathology—A Short Introduction.* (New York: Academic Press, Inc., 1970).

27. Babich, H., and G. Stotzky. "Air Pollution and Microbial Ecology," *Crit. Rev. Environ. Cont.* 4:353–420 (1974).

28. Laurence, J.A. "Effects of Air Pollutants on Plant-Pathogen Interactions," *J. Plant Dis. Protect.* 87:156–172 (1981).

29. Saunders, P.J.W. "Modification of the Leaf Surface and Its Environment by Pollution," in *Ecology of Leaf Surface Microorganisms,* T.F. Preece and C.H. Dickinson, Eds. (New York: Academic Press, 1971), pp. 81–89.

30. Saunders, P.J.W. "Effects of Atmospheric Pollution on Leaf Surface Micro-Flora," *Pestic. Sci.* 4:589–595 (1973).

31. Saunders, P.J.W. "Air Pollutants, Microorganisms and Interaction Phenomena," *Environ. Poll.* 9:85 (1975).

32. Smith, W.H. "Air Pollution—Effects on the Structure and Function of Plant-Surface Microbial-Ecosystems," in *Microbiology of Aerial Plant Surfaces,* C.H. Dickinson, Ed. (New York: Academic Press, 1976), pp. 75–105.

33. Heagle, A.S. "Interactions Between Air Pollutants and Plant Parasites," *Ann. Rev. Phytopathol.* 11:365–388 (1973).

34. Treshow, M. "Interaction of Air Pollutants and Plant Diseases," in *Responses of Plants to Air Pollution,* J.B. Mudd and T.T. Kozlowski, Eds. (New York: Academic Press, 1975), pp. 307–334.

35. Manning, W.J. "Interactions Between Air Pollutants and Fungal, Bacterial and Viral Plant Pathogens," *Environ. Poll.* 9:87–90 (1975).

36. Grzywacz, A., and J. Wazny. "The Impact of Industrial Air Pollutants on the Occurrence of Several Important Pathogenic Fungi of Forest Trees in Poland," *Eur. J. For. Path.* 3:129–141 (1973).

37. Shriner, D.S. "Effects of Simulated Rain Acidified with Sulfuric Acid on Host-Parasite Interactions," PhD Thesis, North Carolina State University, Raleigh, NC (1974).

38. Shriner, D.S. "Effects of Simulated Rain Acidified with Sulfuric Acid on Host-Parasite Interactions," In *First Internat. Symp. on Acid Precipitation and the Forest Ecosystem,* L.S. Dochinger and T.A. Seliga, Eds., USDA For. Serv. Genl. Tech. Rep. No. NE-23, Upper Darby, PA (1975), pp. 919–925.

39. Shriner, D.S. "Effects of Simulated Rain Acidified with Sulfuric Acid on Host-Parasite Interactions," *Water, Air, Soil Poll.* 8:9–14 (1977).

40. Shriner, D.S. "Effects of Simulated Acidic Rain on Host-Parasite Interactions in Plant Diseases," *Phytopathology* 68:213–218 (1978).

41. Evans, L.S., G.R. Hendrey, G.J. Stensland, D.W. Johnson and A.J. Francis. "Acidic Precipitation: Considerations for an Air Quality Standard," *Water, Air, Soil Poll.* 16:469–509 (1981).

42. Cowling, E.B. "Effects of Acid Precipitation and Atmospheric Deposition on Terrestrial Vegetation," in *A National Program for Assessing the Problem of Atmospheric Deposition (Acid Rain),* Galloway et al., Eds., A Report to the Council on Environmental Quality (1978), pp. 47–63.

43. Hutchinson, T.C., and M. Havas. *Effects of Acid Precipitation on Terrestrial Ecosystems* (New York: Plenum Press, 1980).

44. Raynal, D.J., A.L. Leaf, P.D. Manion and C.J.K. Wang. "Actual and Potential Effects of Acid Precipitation on a Forest Ecosystem in the Adirondack Mountains," Publica. No. 56-ES-EHS-79, Research Foundation, State University of New York (1980).

45. Lee, J.J., G.E. Neely, S.C. Perrigan and L.C. Grothaus. "Effect of Simulated Sulfuric Acid Rain on Yield, Growth and Foliar Injury of Several Crops," *Environ. Exp. Bot.* 21:171–185 (1981).

46. Staxäng, B. "Acidification of Bark of Some Deciduous Trees," *Oikos* 20:224–230 (1969).

47. Hoffman, W.A., Jr., S.E. Lindberg and R.R. Turner. "Precipitation Acidity: The Role of the Parent Canopy in Acid Exchange," *J. Environ. Qual.* 9:95–100 (1980).

48. Evans, L.S., and T.M. Curry. "Differential Responses of Plant Foliage to Simulated Acid Rain," *Am. J. Bot.* 66:953–962 (1979).

49. Evans, L.S., N.F. Gmur and F. Da Costa. "Foliar Response of Six Clones of Hybrid Poplar to Simulated Acid Rain," *Phytopathology* 68:847–856 (1978).

50. Evans, L.S., N.F. Gmur and J.J. Kelsch. "Perturbations of Upper Leaf Surface Structures by Acid Rain," *Environ. Exp. Bot.* 17:145–149 (1977).

51. Tukey, H.B. Jr. "Some Effects of Rain and Mist on Plants, With Implications for Acid Precipitation," in *Effects of Acid Precipitation on Terrestrial Ecosystems* T.C. Hutchinson and M. Havas, Eds. (New York: Plenum Press, 1980), pp. 141–150.

52. Myttenaere, C., C. Daoust and P. Roucoux. "Leaching of Technetium from Foliage by Simulated Rain," *Environ. Exp. Bot.* 20:415–419 (1980).

53. Borstrom, F., and G.R. Hendrey. "pH Tolerance of the First Larva Stages of *Lepidurus arcticus* (Pallas) and Adult *Gammarus lacustris,*" G.O. Sars. Report, Zoological Museum, Oslo University, Oslo, Norway (1976).

54. Hagvar, S., G. Abrahamsen and A. Bokhe. "Attack by the Pine Bud Moth in Southern Norway: Possible Effect of Acid Precipitation," *For. Abstr.* 37:394 (1976).

55. Shields, W.J., and S.D. Hobbs. "Soil Nutrient Levels and pH Associated with *Armillariella mellea* on Conifers in Northern Idaho," *Can. J. For. Res.* 9:45–48 (1979).

56. Sierota, Z.H. "Influence of Acidity on the Growth of *Trichoderma viride* Pers. ex Fr. and on the Inhibitory Effect of Its Filtrates Against *Fomes annosus* (Fr.) Cke. in Artificial Cultures," *Eur. J. For. Pathol.* 6:302–311 (1976).

57. Bruck, R.I., S.R. Shafer and A.S. Heagle. "Effects of Simulated Acid Rain on the Development of Fusiform Rust on Loblolly Pine," *Phytopathology* 71:864 (1981).

58. Lacy, G.H., B.I. Chevone and N.P. Cannon. "Effects of Simulated Acidic Precipitation on *Erwinia herbicola* and *Pseudomonas syringae* Populations," *Phytopathology* 71:888 (1981).
59. Berry, F.H., and W. Lantz. "Anthracnose of Eastern Hardwoods," Forest Pest Leaflet No. 133, USDA Forest Service, Washington, DC (1972), p. 6.
60. Hock, W.K., and F.H. Berry. "Sycamore Anthracnose," USDA Agricultural Handbook No. 470 (1975), pp. 88–91.
61. "Forest Insect and Disease Conditions in the United States, 1978," Publica. No. GTR-WO-19, USDA Forest Service, Washington, DC (1980), p. 83.
62. Walton, G.S., and S. Rich. "Serious and Unusual Plant Diseases in Connecticut in 1973," *Plant Dis. Rep.* 58:428–429 (1974).
63. Schuldt, P.H. "Comparison of Anthracnose Fungi on Oak, Sycamore and Other Trees," Boyce Thompson Institute Contributions 18:85–107 (1955).
64. Wiesner, C.J. *Hydrometeorology* (London: Chapman and Hall Ltd., 1970).
65. Likens, G.E., F.H. Bormann, R.S. Pierce, J.S. Eaton and N.H. Johnson. *Biogeochemistry of a Forested Ecosystem* (New York: Springer-Verlag, 1977).
66. Birecka, H., and M.O. Garraway. "Corn Leaf Isoperoxidase Reaction to Mechanical Injury and Infection with *Helminthosporium maydis*. Effects of Cycloheximide," *Plant Physiol.* 61:561–566 (1978).
67. Curtis, C.R., and R.K. Howell. "Increases in Peroxidase Isozyme Activity in Bean Leaves Exposed to Low Doses of Ozone," *Phytopathology* 61:1306–1307 (1971).
68. Horsman, D.C., and A.R. Wellburn. "Synergistic Effect of SO_2 and NO_2 Polluted Air upon Enzyme Activity in Pea Seedlings," *Environ. Poll.* 8:123–133 (1975).

CHAPTER 5

Microbial Response to Acid Deposition and Effects on Plant Productivity

M.K. Firestone
J.G. McColl
K.S. Killham
P.D. Brooks

Plant productivity may be altered profoundly if terrestrial acid deposition affects soil microbial activity. Soil microorganisms are largely responsible for the cycling of nitrogen and sulfur in ecosystems. Since N is the major limiting nutrient in natural terrestrial systems, changes in rates of microbial N transformations could alter the primary productivity of these systems. The major storage form of N is as organic matter, and the nutrient is released for plant utilization through mineralization by microbes. The availability of nitrogen to plants is also a function of the oxidation state; for substantial plant uptake to occur, N must be as NO_3^- or NH_4^+, with specific plants usually preferring one or the other species. Oxidation of NH_4^+ to NO_3^- is catalyzed by soil bacteria during nitrification. Most nitrogen enters the ecosystem through bacterial reduction of N_2 to NH_4^+ (N_2 fixation). Conversely, N is lost from the soil through anaerobic bacterial reduction of NO_3^- to the gaseous species N_2 and N_2O (denitrification).

Among microbial nitrogen transformations, nitrification has been predicted to be one of the processes more sensitive to acid precipitation [1,2] because it is generally believed to be catalyzed by a relatively limited diversity of autotrophic nitrifiers (known to be acid-sensitive on laboratory media). Strayer et al. [3] examined the effects of accelerated acidification on nitrification in surface soil from columns and obtained interesting but complex results. When high NH_4^+ amendments were added to the nitrification assay, all acid treatments tested (pH 3.3–4.1) caused substantial reductions in nitrification rates. However, when NH_4^+ was not added to the soil, the acid treatments generally caused no detectable effect and in some

51

cases caused a slight stimulation in NO_3^- production. These results suggest that in soils with low natural concentrations of NH_4^+ (such as many forest soils) short-term acid deposition may not substantially affect nitrification. The work by Strayer et al. [3] is also consistent with the occurrence of heterotrophic nitrifying organisms in naturally acid forest soils, these heterotrophic nitrifiers being less sensitive to acidity than autotrophic nitrifiers [4,5]. Examining soil exposed to acid rain (pH 3.0) from a coking works for one year, Wainwright [6] found essentially no effect on nitrifying activity.

There has also been concern that symbiotic N_2 fixation may be adversely affected by acid precipitation. Shriner and Johnston [7] found that simulated rain of pH 3.2 applied for 1–9 weeks caused decreased nodulation in kidney beans. Potential effects on a microbial–plant symbiosis can be complex in that they can be mediated through effects on the plant or on the soil.

Of the three microbial nitrogen transformations discussed here, denitrification might be expected to be the least sensitive to acid precipitation simply because it is catalyzed by a relatively diverse group of heterotrophic microbes. While the process is known to be inhibited in extremely acid soils, the overall rates of the reaction have not been found to be particularly sensitive to slight shifts in soil pH [8]. Slight soil acidification does, however, alter the composition of the products formed, causing increased N_2O production relative to N_2 [9].

Distribution of heterotrophic microbial activity in soil reflects the availability of organic carbon. While most types of microbial activity occur to some extent throughout the soil profile, the recognition that maximal activity commonly occurs in somewhat discrete areas of the soil is important to understanding the potential effects of acid deposition. Maximum carbon availability and, hence, maximum heterotrophic activity is usually found near plant roots and in the upper portion of the soil profile, reflecting the site of initial deposition of plant material.

The work reported here focuses on the effects of acid deposition on nitrification, denitrification and N_2 fixation associated with plant roots and nitrification and denitrification in surface soil. We employed simulated acid precipitation of pH 2.0, 3.0, 4.0 and 5.6. The purpose of the extreme treatments (pH 2.0 and 3.0) was not to directly simulate ambient rates but rather to identify thresholds at which effects may begin to occur. We then attempted to assess whether these thresholds may be reached in natural terrestrial systems. We also report results from studies designed to identify physical/chemical mechanisms that may mediate effects of acid deposition on soil microbial activity. The experiments reported here are considered to be preliminary studies intended largely to identify possible "pressure-points" and mechanisms that should be investigated more thoroughly.

APPROACH

Microbial Activity in Rhizosphere Soil

Subterranean clover and Briggs variety barley were grown from seed in a Reiff soil that had been sieved (<4 mm) but not dried before use (Table I). The

Table I. Soil Characteristics

Series	Classification	Texture	pH	% Organic Matter	CEC (meq-100 g^{-1})	Base Saturation (meq-100 g^{-1})
Reiff	Typic xerorthent	Sandy loam	7.4	n.d.	18	18
Shaver	Pachic xerumbrept	Sandy loam	6.4	5.2	19	13
Surnuf (sub surface) sample	Xeric haplohumult	Silt loam	5.3	0.2	27	15
Aiken	Xeric haplohumult	Loam	5.3	3.7	20	12

clover seed was treated with commercial rhizobium inoculant and seeds planted in 6-in.-diameter pots lined with plastic. Plant–soil systems and unplanted soils were spray-irrigated under four treatment regimes (pH 2.0, 3.0, 4.0 and 5.6) at a rate of 250 mL three times per two weeks to total about 13.2 cm of precipitation over ten weeks. The ionic composition of the treatment solution reflected that previously monitored in northern California [10], including a quantity of H_2SO_4 and HNO_3 (2:3 on an equivalent basis) to yield the specified pH plus a standard salt solution consisting of Mg^{2+}, 6; Ca^{2+}, 7; NH_4^+, 15; Cl^-, 15; K^+, 1.5 μeq-L^{-1}. Plants and soils were harvested after 10–12 weeks of treatment for microbial activity assays and nutrient analyses.

Nitrification activity was determined using a modification of the method described by Belser and Mays [11]. Soil samples (5 g) were leached at a constant rate for 7 h on a mechanical extractor with a solution containing NH_4^+ and ClO_3^-. Nitrite production was determined at 2-h intervals using a standard diazotization colorimetric method. Denitrification activity was determined using the acetylene inhibition technique as described by Smith et al. [12]. Nitrous oxide accumulation was quantified by gas chromatography (GC) during incubation periods of less than 8 h. A standard acetylene reduction assay was used to determine the N_2 fixation activity associated with clover roots. The assay was initiated immediately after the root material was harvested and ethylene production was quantified by GC. Soil pH was determined using a 1:1 soil-water slurry.

Four replicate pots were used for each treatment; each pot contained eight plants. Rates of microbial activity were determined by linear regression. When analyses of variance indicated significant differences among treatment means, a Student–Newman–Keuls (SNK) multiple range test was employed [13].

Microbial Activity in Surface Soil

Effects of simulated acid precipitation on activity in surface soil was investigated in five different plant-soil systems. All soils were sieved (<4 mm) and packed in 15-cm-diameter, 30-cm-long plastic cylinders. Shaver soil was planted with soft chess and ponderosa pine seedlings; Reiff soil was planted with clover or barley, or left unplanted. For each plant–soil combination, a set of six replicates was run with and without the addition of fertilizer N and S (N as NH_4NO_3 at 200 kg-ha^{-1}; S as $CaSO_4 \cdot 2H_2O$ at 70 kg-ha^{-1}). The composition of the acid rain, the application rate and the treatment duration were similar to those described in the preceding section. For each plant–soil set, nitrification activity was determined in the top 1 cm of soil. Nitrification and denitrification activities were determined in the surface 1 cm as well as in the bulk soil from which the surface layer had been removed for the Shaver soil containing two-year-old seedlings of ponderosa pine.

Exchangeable NH_4^+ in soil was determined on a 1 N KCl extract using an ammonia electrode. Nitrate was determined by reduction to nitrite with hydrazine sulfate and subsequent diazotization.

Aluminum Mobilization and Toxicity

Five-gram samples of 26 soils were leached with 50 mL of treatment solution over a 2-h period. Treatment solutions of eight pH values were used and the resulting leachates were analyzed for pH and cation content. Acid digests of the leachates were analyzed for Mn and Al by atomic absorption. Leachates from three soils were chosen as being representative of the ranges of Al and Mn content and used for bioassays.

Wells 1-cm in diameter were cored in 1-cm-deep agar plates (Czapek Dox-Oxoid), which were uniformly inoculated with a spore suspension of *Aspergillus flavus*. Soil leachates (0.7 mL) from the pH 2, 3, 4 and 5.5 treatments were added to the center wells. Zones of inhibition of spore germination were first qualitatively documented through visual inspection. Inhibition of spore germination was apparent only from leachates of the pH 2 treatments. To quantify inhibition, the experiment was rerun with the leachates from the pH 2, 3 and 4 inputs and respiration was determined on cores from the plates by GC quantification of CO_2 production. Five replicate flasks were run for each soil-treatment set, each flask containing ten cores (five from each of two plates). For each plate, two sets of cores were tested: one set taken from within the zone of inhibition and one from outside the zone. Bioassays were also run on plates containing the following solutions: 1.0 M acetate–HCl buffer, pH 3.0; buffer (pH 3.0 as before) containing 32 μg-mL^{-1} Al and 20 μg-mL^{-1} Mn; and buffer solutions containing the same concentration of each metal separately. All Al-containing solutions (soil leachates and buffers) were titrated with 1.0 N NaF until free F^- was just detectable by a fluoride-specific electrode. These solutions were also used for parallel assays.

RESULTS AND DISCUSSION

Root-Associated Microbial Activity

Nitrogen fixation activity by clover roots was significantly greater in plants receiving the extreme pH 2 treatment (Table II). This may have been due to the fertilizer

Table II.   Acetylene Reduction by Clover Roots

Input pH	C_2H_4 $(nmol\text{-}g^{-1} - h^{-1})$
2.0	320a[a]
3.0	33.2b
4.0	27.7b
5.6	60.2b

[a] Different letters indicate difference between means at 5%, SNK multiple-range test.

value of the N and S inputs, since in a subsequent experiment in which N and S fertilizers were included there were no significant differences in clover root fixation activity with the different pH inputs.

While there was a pattern of lower nitrification activity with extremely low pH inputs, there were no statistically significant differences (at 5%) associated with input pH in the unplanted or clover rhizosphere soil (Figure 1). In barley rhizosphere soil receiving pH 2 input, nitrification activity was significantly lower. In general, nitrification activity was lower in rhizosphere soil than in unplanted soil and this was significant in two cases (clover, pH 3 and barley, pH 2).

Denitrification activity was significantly higher in barley soil receiving pH 2 treatments but was not significantly different across input pH in clover soil. In the presence of both clover and barley roots, denitrification was substantially greater than that in unplanted soil in which activity was virtually undetectable over the period of assay.

In each plant–soil system the pH of the soil was lower with the extreme acid inputs compared to pH 4.0 and 5.6 inputs. However, the presence of the plant roots also significantly affected the pH of the soil. The presence of clover roots appeared to acidify the rhizosphere soil, and the effect of barley roots was similar. The extent of root influence on soil pH was related to the mass of roots present in the soil. In general, the greater the root mass, the greater was its acidifying effect. Plant mass (root plus shoot weight) was greater in the systems receiving pH inputs of 2 and 3. This increased plant productivity probably resulted largely from the fertilizer value of the acids HNO_3 and H_2SO_4, but also perhaps from increased nutrient availability resulting from slight acidification of the initially neutral soil. Nutrient analysis of the plant material indicated significantly greater S and P concentrations in clover exposed to pH 2 input and greater N concentrations with pH 2 and 3 inputs. Similar patterns were found with barley. Subsequent experiments in which fertilizer N and S were added to the soils substantiate this interpretation of these results (data not shown).

It is well known that plant roots affect the pH of associated soil through differential ion uptake, respiratory CO_2 production and organic acid production [14]. In the experiments reported here, simulated acid precipitation lowered the pH of the bulk soil only when extreme pH 2 treatments were used. However, the presence of plant roots significantly affected the soil pH in five out of eight cases. If one then considers the potentially acidifying effect of realistic quantities of acid precipitation, it seems likely that in rhizosphere soil the influence of plant roots would exceed that of acid precipitation. However, the short-term nature of these experiments should be borne in mind.

The data on microbial activity also reflects the more direct influences of roots on soil microbial processes. Root exudates and sloughed materials provide carbon inputs to root-associated heterotrophs [14]. This can be seen in the high rates of denitrification (a carbon-dependent process) associated with clover and barley roots as compared to unplanted soils. The lower rates of nitrification (an autotrophic process) associated with increased root mass (and increased acid input) is also consistent with a root-originating carbon effect. Thus, through mechanisms

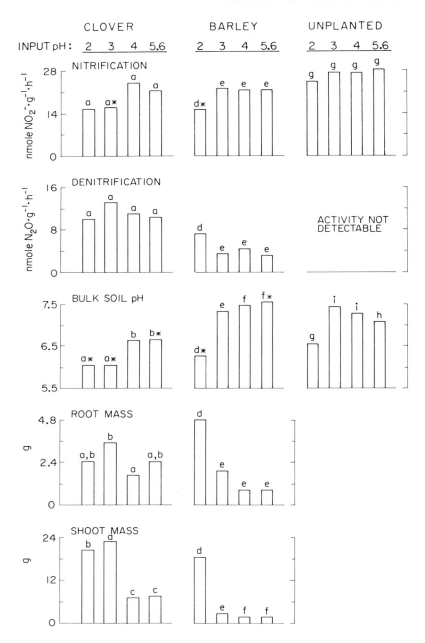

Figure 1. Effects of simulated acid precipitation on microbial activities and plant productivity in a Reiff soil. Within a planting regime, different letters indicate differences between means at 5%, SNK multiple-range test. Differences between a treatment mean in planted soil vs unplanted soil (5%, by Student's t-test) are indicated by.*

of carbon supply, the influence of roots on associated microbial activity may be substantially greater than that of current rates of acid precipitation. Despite the short duration of these exposures, the results reported suggest that microbial activity in rhizosphere soil may be relatively resistant to direct effects of acid precipitation. Unlike rhizosphere soil, the surface soil should be more directly affected by acid precipitation; hence, microbial activity in this zone may be modified.

Microbial Activity in Surface Soil

The effects of simulated acid precipitation on nitrification and denitrification in the surface 1 cm of a Shaver soil containing ponderosa pine seedlings is shown in Fig. 2. It should be noted that the rooting system of a tree seedling is much less extensive than that of a grass or a clover plant; hence, substantial rhizosphere effects are much less likely. In the surface layer of this granitic soil, nitrification activity was significantly depressed with pH 2 treatments, but enhanced in soil from lower in the profile. Determination of exchangeable NH_4^+ indicated that nitrification in surface soil was not inhibited by H^+ mobilization of NH_4^+ cations from the surface soil to lower depths.

The effects on denitrification were almost the reverse of those on nitrification. Denitrification activity was significantly greater in surface soil receiving the extreme input of pH 2. The process of denitrification requires both available NO_3^- and carbon (as well as O_2 limitation). Increased NO_3^- availability resulting from the higher HNO_3 inputs apparently had not stimulated the process in surface soil; nitrate analysis indicated that while the NO_3^- concentration was greater in surface soil receiving pH 2 input, it was similarly greater lower in the profile. Through the use of sterile controls, we also determined that the increased rate of N_2O evolution in soil receiving pH 2 treatment was not due to acid-stimulated chemical decomposition of HNO_2 (data not shown). It is plausible that denitrification activity in surface soil was stimulated by increased carbon availability from soil organic matter which was chemically altered by the pH 2 input and resulting soil acidification. Other work in our laboratory supports this mechanism of effect, but at this time it remains a "working hypothesis."

It is not surprising that the effect of the acid inputs on soil pH was greater in the surface 1 cm than in the bulk soil. The substantial pH decrease (from 6.5 to 3.8) caused by the extreme pH 2 input most likely caused the reduction in nitrification observed in the surface soil. But it should be noted that the less pronounced pH decrease in the lower soil (from 6.5 to 5.4) was not associated with a decrease in nitrifying activity but rather an increase. Interpretation of the nitrification data presented here is facilitated by the results from the eight other plant–soil–fertilizer systems that were studied simultaneously (data not shown). In only one other instance (a fertilized Reiff soil planted in barley) was nitrification activity in surface soil significantly affected by the pH 2 input. In the seven cases in which activity was not significantly altered, the decrease in soil pH resulting from the acid input was much less substantial than that reported here or observed in the

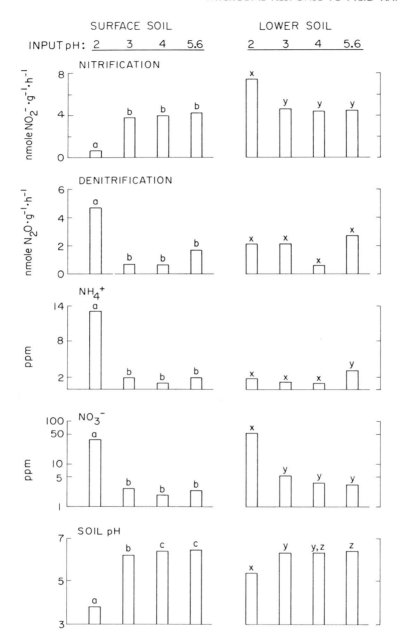

Figure 2. Effects of simulated acid precipitation on nitrification, denitrification and nitrogen species in a Shaver soil planted with ponderosa pine.

Reiff–barley system. Thus, it appears that substantial decreases in pH of surface soil will inhibit nitrification. The critical question then becomes one of the magnitude and rate of soil acidification that may result from realistic or ambient rates of acid precipitation. While the susceptibility of a soil to acidification varies as a relatively complex function of the soil buffering capacity, the degree of acidification caused here by pH 2 treatments should be expected to be uncommon in natural soil systems receiving acid rain.

It is not surprising that acid inputs will affect soil microbial activity (at least on a short-term basis) if there is a substantial decrease in the pH of the soil solution. A somewhat more complex and intriguing question is whether soil microbial activity can be altered by acid precipitation that does not cause a discernible decrease in soil pH. One can hypothesize several mechanisms through which microbes might be affected without a measurable pH shift; two such mechanisms pertain to buffering species in soil. Protonation of soil organic matter functional groups could (perhaps transiently) alter organic matter availability to microbes. Work in our laboratory is addressing this question. Second, aluminum-containing minerals can also buffer soil solution pH. While the mineral species involved are numerous and exact reaction mechanisms complex, the net consumption of H^+ by these minerals usually involves the solubilization or mobilization of $Al^{3+}(OH)_x(H_2O)_y$ forms. The solubilization of Al can be a transient phenomenon with reprecipitation and polymerization reactions rapidly occurring. Can mobilization of aluminum (or manganese) resulting from acid precipitation cause toxicity to soil microbes?

Aluminum Mobilization and Toxicity

The compositions of the leachates used for bioassays are given in Table III. With increasing acidity of the input solution, increasing quantities of Al and Mn were eluted. Between soils, however, the quantity of Al/Mn mobilized was not simply a function of the output solution pH. While the Surnuf soil consistently had the most acid leachate pH, less Al and Mn were eluted than from the two other soils. As previously described in the Approach section, these test solutions were obtained by rapidly leaching a small quantity of soil with a relatively large volume of solution. This is roughly comparable to about 8.9 cm of precipitation input moving through 1 cm of soil in 2 h. The quantities of Al/Mn in the leachates may be unrealistically low due to the short period allowed for Al/Mn solubilization and large volume of solution. Similarly, the leachate period was short for reprecipitation and polymerization reactions to go to completion. However, the question under investigation was one of transient mobilization, not whether Al was mobilized totally out of a soil profile. The problems associated with this type of technique should be borne in mind as the results of the bioassays are discussed.

There were no visually significant effects of leachates from soils treated with pH 5.6, 4.0 or 3.0 solutions on germination of *Aspergillus flavus* spores. Leachates from the Shaver and Aiken soils receiving pH 2.0 inputs did significantly inhibit

Table III.   Aluminum and Manganese Content (μg-mL^{-1} of Leachate) of Soil Leachates

Input pH	Parameter	Surnuf	Shaver[a]	Aiken
			Soil Series	
5.6	Leachate pH	5.0	6.9	5.9
	Al	0.039	0.146	0.140
	Mn	0.001	0.574	0.003
4.0	Leachate pH	4.9	6.8	5.9
	Al	0.010	0.188	0.010
	Mn	0.003	0.644	0.003
3.0	Leachate pH	4.5	6.7	5.8
	Al	0.04	0.12	0.21
	Mn	0.02	2.46	3.70
2.0	Leachate pH	2.7	3.0	3.1
	Al	0.99	32.4	16.1
	Mn	0.16	20.1	25.8

[a] This Shaver series soil is different from that characterized in Table I. This sample had an initial pH of 5.7 and percentage of organic matter of 3.1.

spore germination (Table IV). To determine if unidentified species present in the soil leachates were involved in the observed toxicity, solutions buffered to pH 3.0 and containing Al and Mn concentrations similar to those of the Shaver soil leachates were assayed. While the "leachate simulations" did significantly inhibit spore germination, they were not as toxic as the Shaver and Aiken (high Al/

Table IV.   Effects of Soil Leachates and "Simulated Leachates" on *A. flavus* Spore Germination

Input pH	Surnuf	Shaver	Aiken
		Soil Leached	
4.0	98.8 ± 11.2[a]	108.2 ± 13.6	96.1 ± 8.4
3.0	111.4 ± 14.8	99.6 ± 6.6	100.3 ± 10.1
2.0	115.3 ± 16.5	5.8 ± 2.7	17.7 ± 4.0
2.0 + F$^-$	105.2 ± 12.0	85.4 ± 10.3	92.0 ± 7.9

	pH 3 + Al + Mn	pH 3 + Al	pH 3 + Mn	pH 3
	Simulated Leachates[b]			
Buffer ± Metals	34.4 ± 8.2	39.2 ± 9.5	103.2 ± 8.1	108.0 ± 14.3
Solutions + F$^-$	88.1 ± 6.0	82.6 ± 9.8		

[a] Values given are $100 \times$ [(respiration rates from zones of inhibition)/(respiration rates from unaffected zones of same plates)], ± standard deviation.
[b] 1.0 M acetate-HCl buffer, pH 3, with or without added Al (32 μg-mL^{-1}) and Mn (20 μg-mL^{-1}).

Mn) leachates. This indicates that some unidentified component(s) (perhaps organic ligands) were enhancing leachate toxicity. The Al and Mn were then factored out and the results indicated that the "simulated leachate" toxicity resulted from the Al present in solution. To further substantiate this point, Al-containing solutions were titrated with fluoride as described in the Approach section. When these fluoride-amended leachates and "simulated leachates" were tested, the presence of this strong Al complexing species significantly reduced the solution toxicity. This indicates that a major portion of the leachate effect resulted from Al toxicity.

Solution concentrations of 16 and 32 μg-mL^{-1} Al in leachates from the Aiken and Shaver soils were found to be toxic to *Aspergillus flavus* spore germination; the leachate from the Surnuf soil containing 1 μg-mL^{-1} Al was not toxic. This is consistent with work by Ko and Hora [15], in which a threshold Al toxicity of 0.65 ppm was found for ascospore germination in solution. Other soil microbial processes may be more or less sensitive to Al; fungal spore germination is generally considered to be relatively sensitive to Al toxicity.

The concentrations of Mn found in the soil leachates tested (0.1–26 μg-mL^{-1}) were not found to cause toxicity. Dobereiner [16] reported that additions of 40 ppm Mn to acid soils were inhibitory to symbiotic N_2 fixation. The highest Mn concentration tested in our work (26 μg-mL^{-1}) was lower than that used by Dobereiner.

We have observed microbial Al toxicity only in soil leachates resulting from an extreme acid input of pH 2. However, our data confirm that soils vary substantially in susceptibility for Al and Mn mobilization; the possibility remains that toxic levels of Al may be mobilized by less acidic inputs. Potential toxicity to microbes in terrestrial systems depends on the quantities of Al and/or Mn that may be mobilized by ambient acid precipitation.

CONCLUSIONS

There were no observable effects on the microbial activities tested of the acidified precipitation (pH 4.0) that was most similar to that of "real" acid rain. However, the short durations of the treatments (3 months) precludes statement of direct conclusions concerning effects of long-term, ambient acid deposition. Substantially intensified acid inputs (pH 2 and in some cases pH 3) did affect microbial activity. The total H$^+$ load from three months of pH 2 precipitation would be the same as that from 30 months of pH 3 or 300 months of pH 4. However, the effects that result from pH 2 precipitation cannot be extrapolated to predict long-term effects of more realistic acidified precipitation because the interactions of acidity with the soil physical and microbial components have an extremely important time dependence. Mineral solubilization reactions (involved in buffering), microbial adaptation and reequilibration of organic matter pools are slow processes. Effects which result from accelerated treatments may never occur with less rapid rates of H$^+$ input. But the possibility that changes in microbial activity will result from long-term, "moderate"-pH inputs makes continued investigation, using longer exposure periods, essential.

ACKNOWLEDGMENTS

We thank A. Klein, D. Bush, B. Browne and D. Shelander for their excellent technical assistance. This work was supported in part by a subcontract from the EPA/NCSU Acid Precipitation Program, EPA Cooperative Agreement CR806912 between the U.S. Environmental Protection Agency and North Carolina State University, and by the California Air Resources Board Contract A8–136–31.

REFERENCES

1. Tamm, C.O. "Acid Precipitation: Biological Effects in Soil and on Forest Vegetation," *Ambio* 5:235–238 (1976).
2. Alexander, M. "Effects of Acid Precipitation on Biochemical Activities in Soil," in *Proceedings of an International Conference on Ecological Impact of Acid Precipitation,* D. Drablos and A. Tollan, Eds. (Sandefjord, Norway: SNSF Project, 1980), pp. 47–52.
3. Strayer, R.F., C.J. Lin and M. Alexander. "Effect of Simulated Acid Rain on Nitrification and Nitrogen Mineralization in Forest Soils," *J. Environ. Qual.* 10:547–551 (1981).
4. Johnsrud, S.C. "Heterotrophic Nitrification in Acid Forest Soils," *Holartic Ecol.* 1:27–30 (1978).
5. Remacle, J. "The Role of Heterotrophic Nitrification in Acid Forest Soils—Preliminary Results," *Ecol. Bull.* (Stockholm) 25:561–565 (1977).
6. Wainwright, M. "Effect of Exposure to Atmospheric Pollution on Microbial Activity in Soil," *Plant Soil* 55:199–204 (1980).
7. Shriner, D.S. and J.W. Johnston. "Effects of Simulated, Acid Rain on Nodulation of Leguminous Plants by *Rhizobium* spp.," *Environ. Exp. Bot.* 21:199–209 (1981).
8. Firestone, M.K. "Biological Denitrification," in *Nitrogen in Agricultural Soils,* F.J. Stevenson, Ed. (Madison, WI: American Society of Agronomy, 1982), pp. 289–326.
9. Firestone, M.K., R.B. Firestone and J.M. Tiedje. "Nitrous Oxide from Soil Denitrification: Factors Controlling Its Biological Production," *Science* 208:749–751.
10. McColl, J.G. "A Survey of Acid Precipitation in Northern California," Final Report, California Air Research Board Contract A7–149–30 (1980), pp. 25–51.
11. Belser, L.W., and E.L. Mays. "Specific Inhibition of Nitrite Oxidation by Chlorate and Its Use in Assessing Nitrification in Soils and Sediments," *Appl. Environ. Microbiol.* 39:505–510 (1980).
12. Smith, M.S., M.K. Firestone and J.M. Tiedje. "The Acetylene Inhibition Method for Short-Term Measurement of Soil Denitrification and Its Evaluation Using Nitrogen-13," *Soil Sci. Soc. Am. J.* 42:611–615 (1978).
13. Zar, J.H. *Biostatistical Analysis* (Englewood Cliffs, NJ: Prentice Hall, 1974), p. 151.
14. Russell, E.W. *Soil Conditions and Plant Growth,* 10th ed. (New York: Longman, 1973), p. 552.
15. Ko, W.H., and F.K. Hora. "Identification of an Al Ion as a Soil Fungitoxin," *Soil Sci.* 113:42–45 (1972).
16. Dobereiner, J. "Manganese Toxicity Effects on Nodulation and N_2-Fixation of Beans (*Phaseolus vulgaris* L.) in Acid Soils," *Plant Soil* 24:153–166 (1966).

CHAPTER 6

Biogeochemical Responses of Forest Canopies to Acid Precipitation

Christopher S. Cronan

In recent years, there has been a growing interest in determining how forest canopies interact biogeochemically and physiologically with acidic deposition. Specifically, scientists have been interested in two reciprocal questions: (1) how does acidic deposition affect the chemistry of canopy throughfall and the nutrient status of trees; and (2) what effect does the forest canopy have on the acid-base chemistry of acidic precipitation inputs? In one of the earliest papers on the subject, Eaton et al. [1] reported that approximately 90% of the incoming strong acidity was neutralized as acid precipitation penetrated a northern hardwood forest canopy in central New Hampshire. Similarly, Cole and Johnson [2] found that the canopy of a Douglas fir forest in Washington neutralized most of the incoming acidic precipitation before it reached the forest floor. In Norway, Abrahamsen et al. [3] and Horntvedt et al. [4] reported similar trends for birch trees and also found that increased H^+ ion loading caused increased cation leaching from forest foliage. More recently, several investigators have focused on some of the important mechanisms affecting canopy interactions with acidic deposition. For example, Miller and co-workers [5,6] have tried to separate the relative contributions of canopy washout vs crown leaching in the enrichment of canopy throughfall with various atmospheric pollutants. Meanwhile, Hoffman and co-workers [7,8] have explored the roles of strong and weak acids, ion exchange mechanisms and wet vs dry deposition in modifying precipitation acidity and chemistry.

This chapter presents comparative results that illustrate how forest canopy responses to acidic deposition may differ, depending on forest community species composition, phenological changes and geographic differences in atmospheric loading of strong acids. These results are based on comparative studies of canopy processes in high-elevation coniferous vs lower-elevation hardwood forests of New England. The specific objectives of the investigation were:

1. to use precipitation and throughfall chemistry data to compare the responses of two contrasting forest canopies to regional acid precipitation;

Table I.   Hypothetical Framework for Examining the Effects of Canopy Processes in Coniferous and Hardwood Forests on Solution Chemistry and Acidity of Incident Precipitation

Canopy Process	Effect on pH	Observation
Neutral salt leaching or washout	No net change	Increase in cation and anion equivalents
Canopy washout or leaching of organic or mineral acids	Potential decrease	Increase in total acidity, total anions and perhaps dissolved organic carbon
Uptake of NH_4^+ and release of H^+ by the canopy microflora	Potential decrease	Loss of NH_4^+ and increase in free acidity at constant anionic strength
Ion exchange removal of precipitation-borne H^+	Potential increase	Increase in basic cations correlated with decrease in solution H^+, with no anion increase
Canopy leaching or washout of organic or bicarbonate alkalinity	Potential increase	Increase in cations, decrease in H^+; increase in weak acidity and perhaps alkalinity
Uptake of NO_3^- and release of OH^- or HCO_3^- by the canopy microflora	Potential increase	Loss of NO_3^- and H^+, with no change in major cations or other strong anions

2.   to evaluate several alternative hypotheses concerning the mechanisms by which forest canopies alter acid precipitation chemistry; and

3.   to assess the potential impact of acidic deposition on forest canopy nutrient cycling.

As shown in Table I, a number of focal hypotheses were used to guide this examination of canopy processes. These hypotheses were used in combination with solution chemistry data to elucidate major biogeochemical similarities and contrasts between the hardwood and coniferous canopy subsystems.

METHODS

Field Sites

The comparative field studies were conducted in four sites on Mt. Moosilauke (71°50′ W, 44°1′ N) in the White Mountains of New Hampshire. Mt. Moosilauke (elev. 1462 m) is the westernmost of the White Mountain summits that extend above treeline, and has been the focus of a number of recent studies of ecosystem

Table II. Comparison of Biogeochemical Characteristics for High-Elevation Balsam Fir and Lower-Elevation Northern Hardwood Forests in the White Mountains of Central New Hampshire

Characteristic	Subalpine Zone	Hardwood Zone
Vegetation	Balsam fir	Beech-maple-birch
Soils	Typic cryorthod	Aquic or typic fragiorthod
Leaf Tissue Chemistry[a] (%)		
Ash	2.5	4.9
Calcium	0.21	0.68
Magnesium	0.03	0.17
Potassium	0.46	1.01
Nitrogen	2.35	2.40
Estimated H^+ Loading,		
May–October (meq·m^{-2})	77–100	50–62
Forest Floor Turnover (y)	30	8

[a] Leaf tissue chemistry data for subalpine zone from Lang et al. [13]; data for hardwood zone from Eaton et al. [1] and Gosz et al. [14].

structure and function [9–12]. The two hardwood stands were located at ~520 and 640 m elevation on the west and southeast bases of the mountain. Both stands were midsuccessional northern hardwoods dominated by American beech (*Fagus grandifolia*), sugar maple (*Acer saccharum*), striped maple (*Acer pensylvanicum*) and yellow birch (*Betula alleghaniensis*). The two coniferous stands were located in the subalpine zone at an elevation of 1250–1300 m. These stands were mid- to late-successional age and were dominated by balsam fir (*Abies balsamea*), with a small amount of paper birch (*Betula papyrifera* var. *cordifolia*). The throughfall sampling area in each of the four stands was approximately 0.1 ha.

The rationale for designing a comparative study between the subalpine conifer- ous and northern hardwood forests of New Hampshire was as follows. First, al- though the two systems occur in relatively close proximity to one another, they exhibit several potentially important contrasts in biogeochemical characteristics (Table II) [1,13,14]. Second, previous work had suggested marked differences in throughfall chemistries between the two systems. By comparing throughfall chemis- tries for the two forest types, I hoped to be able to gain additional insight concerning which factors are important determinants of canopy responses to acidic precipita- tion.

Sample Collection and Analysis

Field sampling for this study was conducted during the 1978 growing season (July– October). Bulk precipitation was sampled with two groups of three funnel collectors situated in forest clearings in the hardwood zone at 600 m and in the subalpine coniferous zone at 1250 m. Canopy throughfall was sampled with ten 20-cm-diameter funnel collectors [15] placed randomly in each of the four 0.1-ha sample plots. Water samples were removed from the precipitation and throughfall collectors within 1–3 days of major rain events. During the course of the study, 240 samples of throughfall and 30 samples of bulk precipitation were collected over the course of six rain events. Sample volumes for precipitation collectors ranged from 500 to 1500 mL during those six storms.

Water samples were brought to the laboratory within several hours of collection, and they were usually analyzed the same day for pH (Fisher Accumet 425 equipped with separate pH and reference electrodes). The pH meter was standardized with pH 4.01 and 6.86 buffers and was checked for ionic strength sensitivity using distilled water equilibrated with 320 ppm CO_2 (pH 5.7). Following these measurements, the samples were filtered, stored and analyzed for inorganic constituents as described by Cronan and co-workers [15–17]. In the final data analysis, the solution chemistry data were screened for outlier concentration values before calculation of final averages.

RESULTS

A number of important insights emerged from this comparative analysis of canopy processing in northern hardwood and subalpine coniferous forests. First, there were apparent differences in throughfall volumes between the two systems. In the hardwood stands, throughfall was calculated to be 56 cm for the period May–October (roughly 78% of the bulk precipitation collection). Meanwhile, the fir zone throughfall for the same period was estimated to be 98 cm (approximately 108% of the direct precipitation input to the fir system). These relative and absolute differences in hydrologic fluxes between the two systems may be attributed in part to the following factors: (1) direct precipitation inputs to the high-elevation coniferous system are probably increased by orographic condensation effects; (2) there may be substantial "occult" cloud moisture inputs to the subalpine forest canopy [10]; and (3) the balsam fir forest may experience comparatively lower interceptive-evaporative losses of moisture. In any case, the results showed that the subalpine coniferous forest processes substantially greater quantities of total precipitation than does the lower-elevation deciduous forest. Similar findings have been reported by Reiners and Lang [10] and Lovett [18] who estimated that total precipitation inputs to the high-elevation fir system were $\sim$240 cm-y$^{-1.}$

Comparative Chemistry

The comparative throughfall chemistry results are presented in Figure 1 and Table III. Figure 1 is a representation of the major changes in solution chemistry that occur as similar inputs of acid precipitation pass through the two contrasting forest canopies. In this figure, the upper dark bars show ion concentrations in precipitation entering the two forests, while the open bars show canopy throughfall

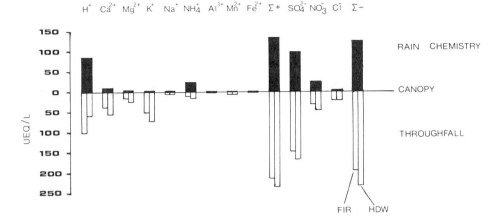

Figure 1. Comparison of precipitation chemistry entering the forest canopy and through-fall chemistry beneath the canopy for a subalpine balsam fir ecosystem and northern hardwood ecosystem in central New Hampshire. Dark bars show concentrations in precipitation; light bars show throughfall concentrations.

concentrations for the fir and hardwood forests, respectively. Several points are evident from these data. First, there is a marked enrichment of most cations and anions as rainfall leaches through both canopies. However, there are differences in the composition and magnitude of these changes. While the balsam fir system may actually show increased hydrogen ion activity in throughfall, the hardwood throughfall generally exhibits less free hydrogen ion and a higher pH than incident precipitation. At the same time, the hardwood throughfall shows greater concentrations of Ca^{2+}, K^+, Mg^{2+}, NH_4^+, SO_4^{2-} and NO_3^- than does the coniferous throughfall.

These patterns are presented more explicitly in Table III, which illustrates the unweighted average throughfall chemistries for the two forest ecosystems. As indicated, calcium and magnesium in coniferous throughfall were enriched by a factor of three over bulk precipitation concentrations. In comparison, hardwood throughfall exhibited a fivefold enrichment factor for calcium and magnesium. Potassium enrichment was even more dramatic for both systems, ranging from a 23-fold increase in coniferous throughfall to a 36-fold increase in hardwood throughfall. Other cations behaved somewhat differently. Sodium concentrations remained virtually unchanged as precipitation moved through both forest canopies. Ammonium ion was distinctive in that its concentration decreased by 50% in the balsam fir canopy and by 30% in the hardwood canopy. Among the trace metals, manganese exhibited the most noticeable enrichment in throughfall for both systems.

As shown in Table III, there were similar enrichment trends for anions in hardwood and subalpine coniferous throughfall. However, the enrichment factors for sulfate, chloride and nitrate were always larger for the northern hardwood system. In terms of electrical charge balance, the hardwood throughfall exhibited

Table III. Comparison of Subalpine Coniferous and Northern Hardwood Forest
Throughfall Chemistries for the 1978 Growing Season[a]

		Sample[b]	
	Bulk Precipitation	Coniferous Throughfall	Hardwood Throughfall
pH	4.06	4.00	4.23
H^+	86	100	59
Ca^{2+}	11	35	56
Mg^{2+}	5	14	23
K^+	2	46	73
Na^+	3	3	3
Al^{3+}	1	2	1
Mn^{2+}	0	5	6
Fe^{2+}	1	2	1
NH_4^+	22	10	15
SO_4^{2-}	98	149	168
NO_3^-	25	30	45
Cl^-	5	17	20
Total Anions	131	217	237
Total Cations	128	196	233
Organic Anions		21[c]	

[a] Concentrations in $\mu eq \cdot L^{-1}$, and represent unweighted means for five storms. Both forest systems receive bulk precipitation chemical inputs with the approximate composition shown above.

[b] Sample means are based on the following collections: July 20 and 31, August 11 and 30, and September 14, 1978.

[c] Organic anions are estimated from the anion charge balance deficit [12].

a virtually perfect anion-cation balance for inorganic ions. Meanwhile, the coniferous throughfall showed an anionic charge deficit that probably represents strongly dissociated organic ligands.

Mechanisms of Neutralization and Acidification

The comparative throughfall data in Table III were also used to evaluate alternative hypotheses to account for the effects of forest canopies on the acid-base chemistry of acidic precipitation inputs. This analysis revealed that there is an apparent net increase in basic cations at the expense of H^+ in hardwood throughfall, and that this increase in Ca^{2+}, Mg^{2+}, K^+ and/or Mn^{2+} occurs without an accompanying additional increase in strong anion equivalents. Thus, compared to input precipitation chemistry, the average hardwood throughfall exhibited an anion increase of 105 $\mu eq \cdot L^{-1}$, a basic cation gain of 133 $\mu eq \cdot L^{-1}$, and a net H^+ loss of 27 $\mu eq \cdot L^{-1}$. While 105 $\mu eq \cdot L^{-1}$ of the anion/cation gain could be attributed to neutral salt leaching and washout from the canopy, the remaining increase in basic cations (28 $\mu eq \cdot L^{-1}$) and decrease in H^+ (27 $\mu eq \cdot L^{-1}$) apparently resulted from a process that occurs under a "fixed anion constraint." As suggested in Table I, this is the type of evidence that would be expected if a leaf surface ion exchange mechanism were responsible for the hardwood canopy neutralization phenomenon. In fact, the hardwood throughfall even exhibited the kind of negative correlation between H^+ and basic cations that one would expect in a simple one-for-one ion exchange mechanism (correlation coefficients for H^+ vs K^+ ranged from -0.62 to -0.96).

However, there is another possible explanation to account for the observed evidence. If the hardwood canopy were to release weak organic or bicarbonate salts to the incoming precipitation, one could have the same increase in K^+, Ca^{2+} or Mn^{2+} concentrations, accompanied by a decrease in H^+ ion activity. In this case, the free hydrogen would be consumed by a weak base and converted to a weak acid. At the same time, one might not detect the weak acid because of its protonated, undissociated form or because of analytical oversight. Thus, the same kinds of chemical clues could result from different neutralization pathways.

In order to examine this weak base buffering hypothesis, several acid-base titrations were performed on samples collected in this study. As shown in Figure 2, the titration results confirmed that the decrease in free acidity in hardwood

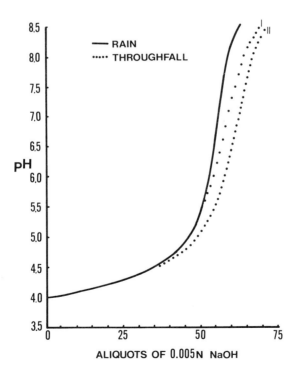

Figure 2. Comparative total acidity titrations for one bulk precipitation and two northern hardwood throughfall samples collected from the same storm in the summer of 1978. Successive 20-μL aliquots of base were added to an initial sample volume of 50 mL that was purged with a nitrogen atmosphere.

throughfall is often accompanied by an increase in weak acidity and alkalinity. For example, throughfall sample II showed 12 μeq-L^{-1} of bound acidity in titrating to pH 6.5 and approximately 16 μeq-L^{-1} of bound acidity in titrating to pH 8.0. These data are interpreted to indicate that soluble weak Brønsted bases may account for an average of 30–50% of the neutralization of incoming strong acids.

Overall, the results indicate that two major neutralization mechanisms may play important roles in the hardwood canopy (Figure 3). These are: (1) canopy ion exchange, in which protons in precipitation exchange with K$^+$ or other cations adsorbed to leaf surfaces; and (2) canopy leaching of weak Brønsted bases, which combine with strong acids from the atmosphere to form weak acids and sulfate and/or nitrate salts. Either of these processes can lead to a net removal of free hydrogen ion from solution and a net leaching of basic cations from the forest canopy.

The coniferous canopy presented an interesting contrast to the hardwood system. Compared to bulk precipitation, the coniferous throughfall exhibited a net acidification that appears to have resulted in part from ammonium uptake in

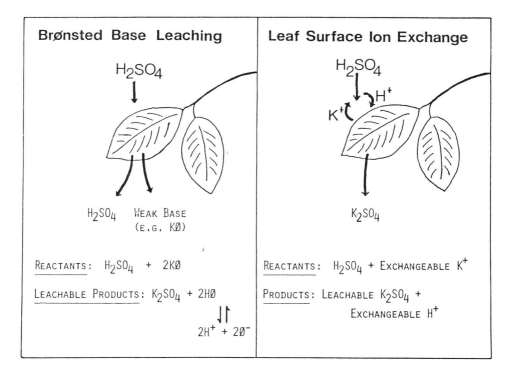

Figure 3.   Postulated neutralization mechanisms for the northern hardwood canopy. The weak Brønsted bases (e.g., KØ) may be either organic or bicarbonate salts of calcium, potassium or manganese. Likewise, the neutral salts may be derived from sulfuric or nitric acids.

the canopy, as well as from washout or leaching of organic or mineral acids. Again, however, this may be only part of the story. Lovett [18] and others have suggested that high-elevation forests of the Northeast receive substantial cloudwater inputs that are two to three times as acid as direct precipitation. Thus, the fir forest throughfall chemistry probably reflects the combined effects of both acidifying and neutralizing processes. In this case, most of the additional free acidity is neutralized and contributes to the neutral salt enrichment in throughfall, while the remaining free acidity contributes to the net acidification of fir zone bulk precipitation inputs.

Comparative Fluxes

The study results also permitted the estimation of water and element fluxes for the hardwood and coniferous forests (Table IV). These fluxes were calculated by estimating hydrologic inputs for precipitation and throughfall (Table IV), and then multiplying these values by the appropriate mean element concentrations shown in Table III. The resulting flux estimates provided several important insights. First, the atmospheric loading rates for H^+ and other ions in wet deposition were shown to be markedly higher for the high-elevation coniferous forest. (These conservative flux estimates would probably increase an additional 33%, if one were to use Lovett's [18] estimates for total precipitation in the subalpine fir zone.) In addition, the comparative throughfall fluxes for the two forest systems were shown to be

Table IV. Comparison of Ion Fluxes for Subalpine Coniferous and Northern Hardwood Forests[a]

Component	Rain		Throughfall	
	Subalpine	Hardwood	Conifer	Hardwood
H_2O^b	90	72	98	56
H^+	77	62	98	33
Ca^{2+}	9.9	7.9	34.3	31.4
Mg^{2+}	4.5	3.6	13.7	12.9
K^+	1.8	1.4	45.1	40.9
Na^+	2.7	2.2	2.9	1.7
NH_4^+	19.8	15.8	9.8	8.4
SO_4^{2-}	88.2	70.6	146.0	94.1
Cl^-	4.5	3.6	16.7	11.2
NO_3^-	22.5	18.0	29.4	25.2

[a] Fluxes are expressed in meq-m^{-2}, and represent estimates for the entire May–October "growing season."

[b] Water flux is expressed in cm. The subalpine direct precipitation volume is based on an estimate by Dingman [19], while other hydrologic fluxes are based on empirical ratios developed from field sampling comparisons between subalpine bulk precipitation, hardwood bulk precipitation, subalpine throughfall and hardwood throughfall.

more similar than one might have guessed from looking at the concentration data in Table III. Although the coniferous throughfall fluxes of H$^+$ and major anions proved to be considerably higher than those in the hardwood system, the fluxes of major cations were estimated to be roughly equivalent between the two systems.

Seasonal Differences in Throughfall Chemistry

The study results also demonstrated the influence of phenological changes on forest canopy responses to acidic precipitation. As shown in Figure 4 and Table V, there were marked seasonal differences in throughfall chemistry for the deciduous hardwood forest. During the growing season, the hardwood throughfall exhibited a mean pH of 4.23 and dominance by strong acids; in contrast, hardwood throughfall collected during the period of senescence preceding leaf-drop showed a mean pH of 5.61 and contained approximately 140 μeq-L^{-1} of alkalinity. By comparison, the coniferous throughfall exhibited very little analogous seasonal change. Overall, the hardwood throughfall data indicated that senescence in the hardwood canopy may lead to dramatic increases in the leaching of alkaline plant metabolites. Thus, during the autumn period of late September to early October, the hardwood canopy may exert an especially profound neutralization effect on atmospheric inputs of acidic precipitation.

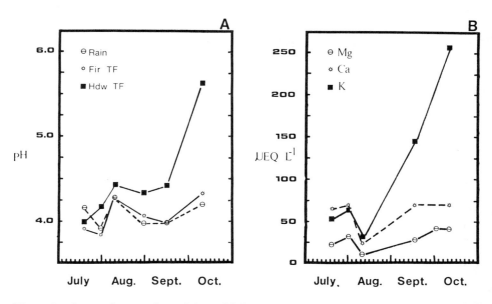

Figure 4. Seasonal comparison of throughfall solution chemistry: (A) comparison of solution pH for bulk precipitation, subalpine coniferous throughfall and northern hardwood throughfall (points represent means of 20 collectors); (B) seasonal trends for calcium, magnesium and potassium in hardwood throughfall.

Table V. Comparison of Northern Hardwood Throughfall Chemistry During the Growing Season vs the Period of Senescence Preceding Leaf Fall[a]

Component	Average Growing Season	October Leaf Senescence
pH	4.23	5.61
H^+	59	2
Ca^{2+}	56	67
Mg^{2+}	23	41
K^+	73	256
Na^+	3	15
NH_4^+	15	2
Al^{3+}	1	
Fe^{2+}	1	
Mn^{2+}	6	
SO_4^{2-}	168	171
Cl^-	20	65
NO_3^-	45	8
Alkalinity	0	139

[a] Concentrations are expressed in $\mu eq \cdot L^{-1}$.

Forest Canopy Nutrient Cycling

The final component of this throughfall study was a comparative analysis of nutrient cycling trends in the coniferous and hardwood forest canopies. The intent here was to compare the cation fluxes in throughfall with the standing crop of each element in hardwood and coniferous foliage. Using fir zone foliage biomass and nutrient data from Cronan [15] and Lang et al. [13], and hardwood foliage biomass and nutrient data from Whittaker et al. [20], Eaton et al. [1] and Gosz et al. [14], estimates were made of the foliage nutrient standing crops for each of the two forest types. Then, the net throughfall fluxes were expressed as a percentage of the foliage elemental standing crop. These boundary calculations indicated that in the hardwood canopy, 20–22% of the calcium and magnesium pools may be cycled during the growing season, whereas roughly 48% of the potassium pool may be turned over through leaching or via the two postulated buffering mechanisms. In comparison, the fir forest data suggested that 23% of the calcium and 37% of the magnesium pools in foliage may be cycled during the growing season, while approximately 37% of the potassium pool may turn over. It should be noted that these values probably represent high estimates, because the net throughfall fluxes include both crown leaching and canopy washout contributions. In any case, these estimates provide a perspective on the magnitude of nutrient cycling in the canopies of these contrasting forest ecosystems, and as such, suggest that there is a considerable neutral salt flux from both types of forest canopies. In terms of mass and charge balance, it should also be noted that uptake and recycling of cations leached from the foliage may result in compensatory proton releases from the root system.

DISCUSSION

This comparative investigation of canopy processing was designed to address a number of hypotheses related to forest canopy interactions with acid precipitation. Results from hardwood and coniferous systems indicated that one of the major processes affecting throughfall chemistry in both forest types is crown leaching and/or canopy washout of neutral salts. While this process may have no net effect on the acid-base chemistry of precipitation inputs, it may result from acid neutralization processes and may represent a significant net removal of nutrient elements from the canopy subsystem. Reports of similar throughfall enrichment responses have been published by other investigators [1,3,21,22].

Besides the common removal of neutral salts from both canopies, there were several distinctive differences in canopy processes between the hardwood and high-elevation systems. The coniferous canopy showed a general tendency toward acidification of bulk precipitation inputs. Similar observations have been reported for other coniferous forest canopies [3,21,23]. This net acidification of precipitation may have resulted from ammonium uptake and corresponding proton release [24] in the coniferous canopy, or may have been caused by canopy washout or leaching of organic or mineral acids. In comparison, the hardwood canopy exerted a neutralizing effect on inputs of acidic precipitation, much in the same way that other hardwood [1] and some coniferous canopies have been shown to respond.

One of the primary aims of this investigation was to examine rigorously the possible neutralization mechanisms operating in the hardwood canopy. Previous throughfall chemistry studies [1,7,8] have emphasized the importance of canopy ion exchange processes as an explanation for forest canopy neutralization of acid precipitation. However, none of these studies presented unequivocal proof that strong acid neutralization could be accounted for by ion exchange. In every case, the authors apparently assumed that the reduction in free acidity was naturally and logically the result of ion exchange. Hoffman et al. [7] even showed the transformation of precipitation strong acidity into throughfall weak acidity and, nevertheless, concluded that their data supported the validity of ion exchange mechanisms. In this investigation, detailed solution chemistry data were combined with testable biogeochemical hypotheses to estimate the relative roles of different neutralization mechanisms in the northern hardwoods canopy. These estimates indicated that, during the growing season, approximately 30–50% of the strong acid neutralization in the hardwood canopy may be accounted for by weak Brønsted base leaching (e.g., $KHCO_3$ or $K_2C_2O_4$). The remaining neutralization is attributable to leaf surface ion exchange. In contrast, during the period of senescence preceding leaf-drop, most of the hardwood canopy neutralization apparently results from leaching of alkaline metabolite salts from the hardwood leaves.

Based on the current literature, it appears that additional studies are needed to examine the potential impacts of acidic deposition on plant physiological processes in forest species. At this time, it is not known to what extent acidic precipitation may influence the electrical charge and hydrophobicity of leaf surfaces, the pH-related control of stomatal movements, the pH-dependent activity of plant enzymes, or the overall phenology of leaf senescence.

CONCLUSIONS

The data from this comparative study were interpreted in relation to the hypotheses outlined in Table I. The resulting conclusions were as follows:

1. In both forest systems, canopy throughfall exhibits marked ionic enrichment, apparently derived from neutral salt leaching or washout.
2. Direct precipitation inputs undergo virtually no neutralization in transit through the balsam fir forest canopy. In fact, fir zone direct precipitation may actually be acidified by one of two likely mechanisms: canopy uptake of NH_4^+ and associated H^+ release; or canopy leaching or washout of organic or mineral acids accumulated on the foliage. Overall, the fir forest canopy may neutralize some fraction of the total wet and dry input of strong acids; yet, the net effect of the coniferous forest canopy is to acidify the solution chemistry of direct precipitation.
3. Acid precipitation may be neutralized to a noticeable degree by the hardwood forest canopy. In terms of seasonal trends, this neutralization phenomenon becomes particularly dramatic during leaf senescence.
4. Canopy neutralization of acidic deposition in hardwood and coniferous forests occurs through two mechanisms: washout or leaching of weak Brønsted bases from foliar surfaces, or leaf surface ion exchange removal of H^+ ions.
5. Total precipitation and H^+ ion loading may vary substantially between different elevations in the Northeast. Although throughfall concentrations in high-elevation coniferous forests may be lower than in northern hardwood forests, the coniferous system processes larger amounts of total precipitation and may consequently exhibit net throughfall element fluxes that are similar to fluxes in the lower-elevation hardwood forest.

ACKNOWLEDGMENTS

The research reported here was sponsored by U.S. DOE contracts EV04498 and EV10750.

REFERENCES

1. Eaton, J.S., G.E. Likens and F.H. Bormann. "Throughfall and Stemflow Chemistry in a Northern Hardwood Forest," *J. Ecol.* 61:495–508 (1973).

2. Cole, D.W., and D.W. Johnson. "Atmospheric Sulfate Additions and Cation Leaching in a Douglas Fir Ecosystem," *Water Resources Res.* 13:313–317 (1977).

3. Abrahamsen, G., K. Bjor, R. Horntvedt and B. Tveite. In: *Impact of Acid Precipitation on Forest and Freshwater Ecosystems in Norway,* F.H. Braekke, Ed. (Oslo-As, Norway: SNSF Project, 1976), pp. 38–63.

4. Horntvedt, R., G.J. Dollard and E. Joranger. In: *Ecological Impact of Acid Precipitation,* D. Drabløs and A. Tollan, Eds. (Oslo-As, Norway: SNSF Project, 1980), p. 192.

5. Lakhani, K.H., and H.G. Miller. In: *Effects of Acid Precipitation on Terrestrial Ecosystems,* T.C. Hutchinson and M. Havas, Eds. (New York: Plenum Press, 1980), p. 161.

6. Miller, H.G., and J.D. Miller. In: *Ecological Impact of Acid Precipitation,* D. Drabløs and A. Tollan, Eds. (Oslo-As, Norway: SNSF Project, 1980), p. 33.

7. Hoffman, W.A., S.E. Lindberg and R.R. Turner. "Precipitation Acidity: The Role of the Forest Canopy in Acid Exchange," *J. Environ. Qual.* 9:95–100 (1980).

8. Lindberg, S.E., D.S. Shriner and W.A. Hoffman, Jr. "The Interaction of Wet and Dry Deposition with the Forest Canopy," in *Acid Precipitation: Effects on Ecological Systems,* F.M. D'Itri, Ed. (Ann Arbor, MI: Ann Arbor Science Publishers, 1982), pp. 385–410.

9. Vitousek, P.M. "The Regulation of Element Concentrations in Mountain Streams in the Northeastern United States," *Ecol. Monog.* 47:67–87 (1977).

10. Reiners, W.A., and G.E. Lang. "Vegetation Patterns and Processes in the Balsam Fir Zone, White Mountains, New Hampshire," *Ecology* 60:403–417 (1979).

11. Schlesinger, W.H., and W.A. Reiners. "Deposition of Water and Cations on Artificial Foliar Collectors in Fir Krummholz of New England Mountains," *Ecology* 55:378–386 (1974).

12. Cronan, C.S. "Solution Chemistry of a New Hampshire Subalpine Ecosystem: A Biogeochemical Analysis," *Oikos* 34:272–281 (1980).

13. Lang, G.E., W.A. Reiners and G.A. Shellito. "Tissue Chemistry of *Abies balsamea* and *Betula papyrifera* var *cordifolia* from Subalpine Forests of Northeastern U.S.A.," *Can. J. For. Res.* (in press).

14. Gosz, J.R., G.E. Likens and F.H. Bormann. "Nutrient Content of Litter Fall on the Hubbard Brook Experimental Forest, New Hampshire," *Ecology* 53:769–784 (1972).

15. Cronan, C.S. "Solution Chemistry of a New Hampshire Subalpine Ecosystem: Biogeochemical Patterns and Processes," PhD Thesis, Dartmouth College (1978).

16. Cronan, C.S. "Determination of Sulfate in Organically Colored Water Samples," *Anal. Chem.* 51:1333–1335 (1979).

17. Cronan, C.S., W.A. Reiners, R.C. Reynolds and G.E. Lang. "Forest Floor Leaching: Contributions from Mineral, Organic, and Carbonic Acids in New Hampshire Subalpine Forests," *Science* 200:309–311 (1978).

18. Lovett, G.M. "Forest Structure and Atmospheric Interactions: Predictive Models for Subalpine Balsam Fir Forests," PhD Thesis, Dartmouth College (1981).

19. Dingman, S.L. "Water Balance as a Function of Elevation in New Hampshire and Vermont," unpublished results (1978).

20. Whittaker, R.H., F.H. Bormann, G.E. Likens and T.G. Siccama. "The Hubbard Brook Ecosystem Study: Forest Biomass and Production," *Ecol. Monog.* 44:233–254 (1974).

21. Feller, M.C. "Nutrient Movement Through Western Hemlock–Western Red Cedar Ecosystems in Southwestern British Columbia," *Ecology* 58:1269–1283 (1977).

22. Henderson, G.S., W.F. Harris, D.E. Todd, Jr. and T. Grizzard. "Quantity and Chemis-

try of Throughfall as Influenced by Forest-Type and Season," *J. Ecol.* 65:365–374 (1977).

23. Johnson, D.W. "Processes of Elemental Transfer in Some Tropical, Temperate, Alpine, and Northern Forest Soils: Factors Influencing the Availability and Mobility of Major Leaching Agents," PhD Thesis, University of Washington (1975).

24. Lang, G.E., W.A. Reiners and R.K. Heier. "Potential Alteration of Precipitation Chemistry by Epiphytic Lichens," *Oecologia* 25:229–241 (1976).

CHAPTER 7

Assessing the Possibility of a Link Between Acid Precipitation and Decreased Growth Rates of Trees in the Northeastern United States

Arthur H. Johnson
Thomas G. Siccama
Robert S. Turner
Deborah G. Lord

Recent experiments indicate that acid precipitation might adversely affect trees directly [1–3] or possibly indirectly through unfavorable effects on soils [4–6]. Field surveys of tree growth have suggested recent declines in growth rate [7–9], but in most cases there is no information regarding the cause(s).

This chapter presents evidence from tree ring studies and pertinent biogeochemical data. Clear, widespread and sustained decreases in tree growth rate beginning in the last 2–3 decades are documented. Pitch pine (*Pinus rigida*) and shortleaf pine (*Pinus echinata*) are affected in southern New Jersey, and red spruce in Vermont. Growth of pitch and shortleaf pine in New Jersey, and pitch pine, white pine (*Pinus strobus*) and chestnut oak (*Quercus pinus*) in the Shawangunk Mountains of New York [10] show an increase in the sensitivity of growth to temperature and moisture beginning in the 1950s. Thus, the appearance or strengthening of some regional-scale stress is suggested. Focusing solely on acid precipitation as a possible cause would be restrictive, since there is a possibility of combined or synergistic effects of the many biotic and abiotic stresses acting on forests.

The stress(es) affecting red spruce are causing a widespread dieback. That case differs from the case in the Middle Atlantic states, where the trees are not dying. Consequently, the causes may be markedly different. It is of paramount importance to assess as many potential causes as possible by testing as many hypotheses as necessary.

METHODS

The procedures for tree ring collection and analysis are given by Johnson et al. [9]. Dominant and open grown trees were selected to limit the effects of competition. Soil analyses were carried out using standard procedures [11,12]. Analyses of water samples were done according to U.S. Geological Survey (USGS) procedures. Analysis of plant tissue was accomplished by dry ashing [11] followed by plasma emission spectroscopy.

RESULTS AND DISCUSSION

New Jersey Pinelands study

The Pinelands region of southern New Jersey encompasses approximately 2500 km². The growth rate decline is most pronounced on the deep, sandy upland soils, which are very acid (pH 3.4–4.6) and have a low cation exchange capacity (0–4 meq-100 g⁻¹). There is relatively little neutralization by ion exchange or mineral weathering as precipitation moves through the soil [13]. Long-term sampling of two headwater streams by USGS indicates acidification of the forest ecosystem. Stream pH declined significantly between 1963 and 1978 [13,14] (Figure 1). Anion chemistry and electroneutrality balances showed that H_2SO_4 is the dominant source of free H^+. During 1978–1981, stream pH increased apparently in response

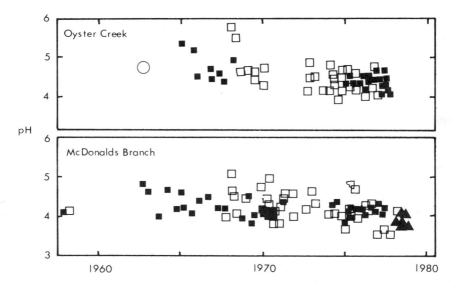

Figure 1.   Stream pH at McDonalds Branch and Oyster Creek, New Jersey. Different symbols represent measurements from different studies or data of questionable quality [13,14]. The pH decreases are statistically significant (modified from Johnson [13]).

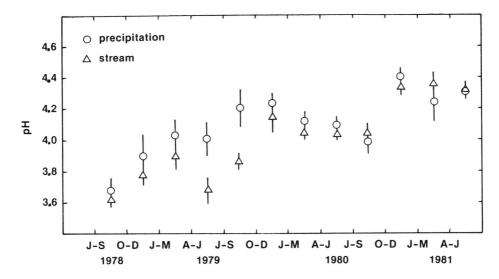

Figure 2. Relationship between precipitation and stream pH (after Johnson et al. [9]). Symbols represent seasonal averages ± one s.e. of the mean. The data summarize 82 collections. Stream values are lagged one season (three months) behind precipitation values.

to decreasingly acid precipitation (Figure 2). Several mechanisms were analyzed that could have accounted for the acidification between 1963 and 1978; the most reasonable explanation is acidification of the precipitation, coupled with a lack of buffering by the old, highly weathered soils [13,14].

About one-third of the trees growing on the well drained upland soils (Quartzipsamments and Hapludults) showed a readily apparent, abnormal decline in increment size. Abnormally narrow rings first appeared between 1955 and 1965 regardless of age or species. The first year of diminished growth varied from tree to tree, with 1957 or 1958 being the most common starting point. The decreased growth effect is clearly visible in old and young trees, in planted and native stands of pitch and shortleaf pine. Figure 3 summarizes the data for three groups of native trees. There was a sharp decline in growth rate during the 1950s with no subsequent recovery. The 40 to 60 y class and the 80 to 100 y class grew at similar rates initially, but after 20 years, the growth rates differed substantially. The younger class grew about two-thirds as fast from age 20 to 45. The period of reduced growth of the younger trees was 1955–1979, a period during which precipitation acidity is thought to have intensified and spread [15]. We assume that the older trees grew "normally" from 1890 to 1950 and are therefore a suitable "control" population with which recent growth can be compared meaningfully. All of the plausible causes were investigated for which information could be found. Symptoms resulting from controlled burning, wildfire, drought, pests and disease were absent or limited to one or two areas.

Comparison of the patterns of stream pH and growth rates showed a marked similarity during the 1963–1979 period (Figure 4). Since there are only a few

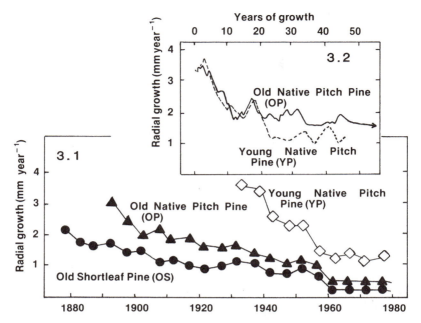

Figure 3.   Radial growth vs time for three groups of mature trees in the New Jersey Pinelands (after Johnson et al. [9]). For OP, n = 30; YP, n = 32; OS, n = 15. Trees were collected from six sites. Symbols in 3.1 represent five year averages. 3.2 shows each year's growth as a function of age. In 3.2, OP and YP are aligned to show the first 45 years of growth.

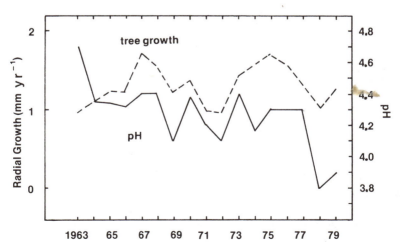

Figure 4.   Mean annual stream pH at McDonalds Branch and mean yearly increment size of young native pitch pine (YP in figure 3) (modified from Johnson et al. [9]).

New Jersey pinelands.

years of precipitation pH data for the Pinelands, we tested the relationship of stream pH to growth rates. Figure 2 suggests that stream pH reflected the pH to which trees and soils in the region were subjected, particularly since Oyster Creek and McDonalds Branch showed similar year-to-year variations in pH during the 1960s and 1970s [14]. We tested the relationships between growth rates and the environmental variables shown in Table I using stepwise and all-possible-combinations multiple regression techniques. The results show that before 1955, neither of the environmental variables (temperature and moisture) showed a consistently significant relationship to growth. During the last 17 years of record, pH was related to growth in all cases. Winter moisture, winter temperature, spring and summer insolation, and summer drought frequently appeared as significant variables and some effect on growth is suggested.

Puckett [10] analyzed climatic trends in Mohonk Lake, New York, and growth rates of pitch pine, white pine, chestnut oak and eastern hemlock (*Tsuga canadensis*) from 1900 to 1973. In 1954–1973 the response of pitch and white pine and chestnut oak growth to variations in temperature and precipitation was significantly greater than in the other periods tested (1901–1920 and 1926–1945). Consideration of other stresses (e.g., gypsy moth infestation, drought and climatic change) left some response to air pollution as a plausible explanation.

The change in the factors to which tree growth responded and the decreased growth rate in southern New Jersey suggest the emergence or strengthening of a stress during the 1950s. The Pinelands tree ring record dates back to 1852, and no other recorded events are as severe, widespread and long-lasting. The evidence linking acid precipitation to the decreased growth rates and shift in growth determining factors is purely circumstantial, based on the timing of the changes and correlation with changes in stream pH. There is, however, no evidence supporting any other mechanisms, and there is some indication that neither SO_2 nor ozone is a major factor. The response of white pine in the Pinelands is different from that of the other species tested. Two plantations (23 trees) were sampled. The trees were 40–45 years old, and showed no noticeable decline in growth rate [9], nor was growth related to pH or drought (Table 1). As white pine is sensitive to SO_2, it is difficult to argue in support of SO_2 as the dominant stress. Average growing season ozone concentration and the frequency of high ozone events during the growing season are positively (occasionally significantly) correlated with growth. Although the period of record is short (1973–1979), it is difficult to argue that ozone is a major factor limiting growth.

Owing to the political and economic implications of linking tree growth to acid precipitation, it is important to demonstrate a cause-and-effect relationship if it exists. Experimental work, of course, offers the best opportunity.

Green Mountains Study

Very substantial mortality of red spruce in all size classes has been documented in the northern Green Mountains, amounting to a 50% decline in density and

Table I. Summary of Multiple Regression Analyses[a]

Independent Variables	White Pine (n = 23)[c]	Young Pitch (n = 32)	Plantation Pitch (n = 9)	Plantation Lob. (n = 29)	Plantation Shortleaf (n = 38)	Old Shortleaf (n = 15)		Old Pitch (n = 30)	
						(1930–1954)	(1963–1979)	(1930–1954)	(1963–1979)
Summer Temperature									
Spring Temperature						−,1			
Winter Temperature		+,2		+,2	+,2				
Fall Temperature									
Summer Precipitation									
Spring Precipitation									
Fall Precipitation									
Winter Precipitation	−,2		−,2						
Summer Sunshine	+,1			+,1	+,1	nd		nd	
Spring Sunshine		+,1				nd		nd	
Previous Year Growth				+,1		nd	+,1	nd	
Yr. Avg. O₃ Conc.						nd		nd	
Summer O₃ Conc.						nd		nd	
Summer Drought Index	−,3	−,3		−,3	−,3	nd	−,3	nd	−,3
pH	+,1	+,1	+,1	+,2	+,2	nd	+,2	nd	+,1

[a] Mean yearly increment size or mean growth index value was used in conjunction with the independent variables.
[b] + = positive correlation; − = negative correlation; 1 = significant at 5% level in some combinations; 2 = significant at 5% level in most combinations; 3 = significant at 5% level in combination with pH; nd = no data for independent variable.
[c] n is the number of tree cores in each group.

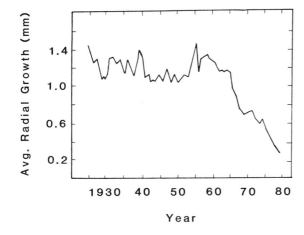

Figure 5. Growth rate of *Picea rubens* in the boreal zone on Camels Hump, Vermont. Line represents the mean of the 10 oldest trees (100–220 y) cored in 1980.

basal area between 1965 and 1980 [16]. The cause is unknown, but appears to be unrelated to successional dynamics [16] or to climate [17]. Field observations indicate spruce mortality in the Adirondacks of New York, the Green Mountains of southern Vermont and in the White Mountains of New Hampshire. In the northern Green Mountains, the most obvious symptom is needle loss beginning in the crown and eventually progressing downward through the lower branches. The tree ring record as reflected in Figure 5 suggests that this takes several years in mature trees. Reports of similar patterns of needle loss from Norway spruce in Germany [18] and Finland [19], where acid deposition and air pollutants are suspected causes, suggested the possibility of an abiotic stress.

We present here the initial findings on soil, foliar and root chemistry. Work on physiological parameters is in progress. Spruce has died out in the three major forest types of northern Vermont—the northern hardwood forest (NHW), where sugar maple (*Acer saccharum*), American beech (*Fagus grandifolia*) and yellow birch (*Betula alleghaniensis*) are dominant; the high-elevation boreal forest (BOR), where balsam fir (*Abies balsamea*) red spruce and white birch (*Betula papyrifera* var. cordifolia) are dominant; and the intervening transition forest (TRNS).

As Camels Hump Mountain (Huntington, Vermont) provided a variety of soils and forest types in which spruce are dying, and as there were baseline data on soil chemistry from 1965 [12], the composition of soils, fine roots and foilage were studied to see if there was any evidence that suggested acid rain as a contributing stress. Geologic and soil characteristics are explained in detail elsewhere [12]. In June 1981 red spruce approximately 1 m tall, which were 20–30 years old, and trees in the 2 to 9 cm size class which were 30–120 years old were sampled. Trees were judged to be in good condition if they had foliage in the crown and buds. Trees were judged to be in poor condition if they had no needles in the crown. Trees in poor condition had few or no buds. As a "control" site, we found a healthy stand of small spruce in the Hubbard Brook Experimental Forest (West Thornton, New Hampshire). It should be noted that there is some spruce mortality

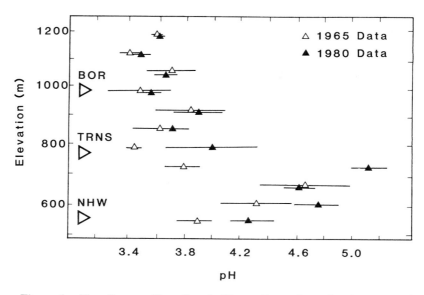

Figure 6.   Elevation vs pH at Camels Hump. Darts show the elevations at which spruce were sampled. Bars represent ± one s.e. of the mean. In 1965 three samples per site were analyzed, and in 1980 five samples per site were analyzed.

in other stands at Hubbard Brook. The stand sampled was at approximately 800 m above sea level, a transition forest zone.

At Camels Hump, soil pH in the O horizons (which is the root zone for <2-cm trees) did not change between 1965 and 1980 (Figure 6). Fine root (i.e., feeder roots <1 mm diameter) composition is shown in Table II and significant differences are summarized in Table III. Owing to the rather large difference in root Al levels between control and affected areas, we sampled healthy looking spruce in other areas to assess root Al levels (Table IV). Particular interest in root Al was prompted by the work of Ulrich [5,6], which suggests a link between Norway spruce mortality and high concentrations of Al in soil moisture and root tissue. In the mountains of northern New England, the authors' data suggest that trees in good condition can have high (>2000 ppm) concentrations, and trees in poor condition (i.e., BOR <2 cm, Table II) can have low (<500 ppm) Al concentrations.

In light of the fact that spruce growth continued to decline during the period when soil pH did not change, and given the general lack of significant differences in root composition among the groups of trees tested, there is no compelling evidence that acid precipitation affected red spruce through a soil pathway. The evidence does not support the contention that Al is an important link between acid precipitation and spruce growth/mortality. The cause and/or effect of the elevated Zn and Pb levels at Camels Hump are presently unknown.

Results of foliar analyses are shown in Table V and significant differences

Table II. Fine Root Composition (*Picea rubens*)

Forest Type	Condition	N (%)	S (%)	P (%)	K (%)	Ca (%)	Mg (%)	Mn (ppm)	Cu (ppm)	Zn (ppm)	Al (ppm)	Pb (ppm)	Cd (ppm)	Ni (ppm)	Cr (ppm)	Mo (ppm)
2–9 cm DBH																
Camels Hump																
TRNS (n = 10)	Poor	1.1	0.09	0.11	0.27	0.19	0.07	280	15	290	4800	65	7	7	6	14
	Good	1.1	0.08	0.10	0.24	0.24	0.06	280	11	220	3100	82	5	7	5	<5
BOR (n = 20)	Poor	0.91	0.07	0.08	0.25	0.22	0.06	480	8	180	1300	47	<5	<5	<5	<5
	Good	0.98	0.08	0.10	0.25	0.22	0.07	200	7	170	1700	48	<5	<5	<5	<5
<2.0 cm DBH																
Camels Hump																
NHW (n = 10)	Poor	0.61	0.07	0.07	0.30	0.50	0.10	350	9	170	3500	25	<5	12	<5	<5
	Good	0.72	0.06	0.07	0.31	0.48	0.14	250	12	190	2800	30	<5	13	<5	<5
TRNS (n = 20)	Poor	1.0	0.07	0.09	0.31	0.29	0.08	350	9	220	1300	45	6	6	<5	<5
	Good	1.1	0.08	0.11	0.36	0.27	0.09	280	0	260	1200	50	6	6	<5	<5
BOR (n = 20)	Poor	nd[a]	nd	0.07	0.24	0.29	0.08	170	9	210	490	60	5	<5	<5	<5
	Good	nd	nd	0.10	0.27	0.24	0.08	230	11	190	1400	65	<5	<5	<5	<5
Hubbard Brook[b] (n = 11)		0.75	0.06	0.08	0.42	0.25	0.08	190	7	98	410	31	<5	<5	<5	<5

[a] nd = no data.
[b] No obvious mortality.

Table III.   Comparisons Among Root Compositions

Significant differences in root composition between Camels Hump and Hubbard Brook:

1. more K at Hubbard Brook (1.5X);
2. more Zn and Pb at Camels Hump (2X); and
3. more Al at Camels Hump (5X).

Significant trends in root composition at Camels Hump (<2 cm class):

1. Pb directly related to elevation; and
2. Al inversely related to elevation.

No significant differences in root composition between trees in good vs poor condition at Camels Hump.

Table IV.   Al Concentrations in *Picea rubens* Fine Roots (Trees <2 cm diameter)

Location	Mortality	Al (ppm)	n
Camels Hump			
NHW	Yes	3130 ± 760	10
TRNS	Yes	1260 ± 202	20
BOR	Yes	957 ± 356	20
Overall Mean		1513 ± 225	
Hubbard Brook			
NHW	No	410 ± 228	11
Road Cut	No	2144 ± 208	10
Lincoln, VT, Old Field	No	2063 ± 176	10
Bristol, VT, Undisturbed	No	821 ± 62	8
Manchester, VT[a]			
Old Field A	No	762 ± 131	4
Old Field B	No	729 ± 104	6
Old Field C	No	497 ± 64	5
Overall Mean		1168 ± 167	

[a] Canopy trees in the Manchester old field stands were dead or dying, and were infected with *Armillaria*.

are summarized in Table VI. Foliar N and S levels are directly related to elevation (Figure 7) and foliar S and S:N ratios are higher in trees that are in poor condition (Figures 7 and 8). There has been little work published on the effects of acid rain on red spruce, and meaningful comparisons with other species are difficult. It is interesting to note, however, that S levels at Camels Hump in poor trees are in the range associated with physiological stress in Scotch pine (*Pinus sylvestris*) [20] and mesophyll cell damage in Norway spruce [21] if the excess S is the result of SO_2. On the other hand, when S in excess of protein requirements is delivered

Table V. Foliage Composition (*Picea rubens*)

Forest Type	Condition	N (%)	S (%)	P (%)	K (%)	Ca (%)	Mg (%)	Mn (ppm)	Cu (ppm)	Zn (ppm)	Al (ppm)	Pb (ppm)	Cd (ppm)	Ni (ppm)	Cr (ppm)	Mo (ppm)
2–9 cm DBH																
Camels Hump																
TRNS (n = 10)	Poor	1.1	0.10	0.12	0.48	0.22	0.05	347	3	15	60	3	<3	<3	<3	<3
	Good	1.3	0.09	0.19	0.56	0.29	0.08	565	6	23	39	<3	<3	<3	<3	<3
BOR (n = 20)	Poor	1.0	0.12	0.12	0.46	0.31	0.08	500	<3	23	35	<3	<3	3	<3	<3
	Good	1.2	0.10	0.16	0.48	0.23	0.08	400	<3	19	40	3	<3	4	<3	<3
<2 cm DBH																
Camels Hump																
NHW (n = 10)	Poor	0.89	0.08	0.10	0.50	0.30	0.06	220	<3	23	52	<3	<3	3	<3	5
	Good	0.89	0.07	0.10	0.48	0.29	0.06	250	3	17	43	<3	<3	3	<3	<3
TRNS (n = 20)	Poor	1.0	0.10	0.10	0.45	0.28	0.05	420	<3	19	28	<3	<3	<3	<3	<3
	Good	1.0	0.08	0.10	0.43	0.26	0.05	410	<3	20	32	<3	<3	<3	<3	<3
BOR (n = 20)	Poor	1.1	0.12	0.10	0.40	0.44	0.09	400	<3	30	32	3	<3	<3	<3	<3
	Good	1.1	0.10	0.10	0.39	0.34	0.07	340	3	26	63	<3	<3	<3	<3	<3
Hubbard Brook[a] (N = 11)		0.99	0.08	0.09	0.56	0.32	0.06	250	4	21	23	<3	<3	<3	<3	<3

[a] No obvious mortality.

Table VI.   Comparisons Among Foliar Compositions

Significant differences in foliar composition between Camels Hump and Hubbard Brook:

1. more K at Hubbard Brook;
2. more S at Camels Hump; and
3. more Al at Camels Hump.

Significant trends in foliar composition at Camels Hump (<2 cm size class):

1. N increases with elevation;
2. S increases with elevation; and
3. K is inversely related to elevation.

Significant differences in foliar composition between trees in good vs poor condition at Camels Hump:

1. higher S in poor trees; and
2. higher S:N ratio in poor trees.

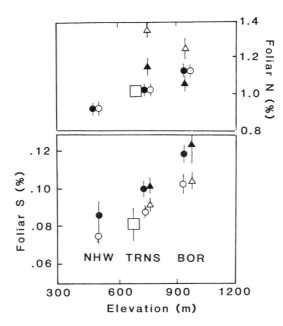

Figure 7.   Foliar N and S in *Picea rubens.* Circles are <2-cm trees at Camels Hump, triangles are 2 to 9-cm trees, and squares are <2-cm trees at Hubbard Brook. Solid symbols are trees in poor condition. Open symbols are trees in good condition.

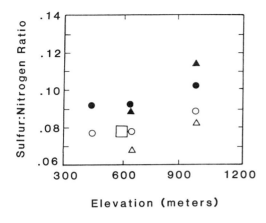

Sulfur-to-nitrogen ratios in *Picea rubens* needles. Symbols are the same as in Figure 7.

as artificial acid rain, Norway spruce and Scotch pine accumulate the same levels of foliar S without apparent harm [22].

SO₂ measurements at Hubbard Brook (approximately 100 km east of Camels Hump) and at Whiteface Mountain (approximately 100 km west) are very low, averaging 2.5 μg m^{-3} in 1973–1975 [23]. This is about an order of magnitude less than levels thought to cause damage through long-term exposure [24].

There is a trend in foliar K which parallels the trend in O-horizon exchangeable K (Figure 9). Foliar K levels at Camels Hump appear to be low compared

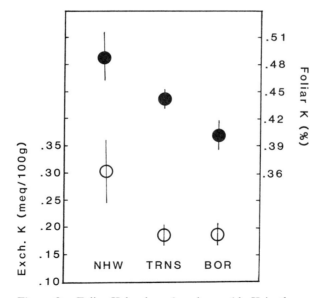

Figure 9. Foliar K levels and exchangeable K in the forest floor. Foliar K data are from Table II. Exchangeable K data are from Siccama [12].

to Hubbard Brook, and there is a possibility that foliar leaching contributed to the low K levels, but there is at present no evidence linking the K differences to acid precipitation or to the dieback.

Initial reports on the spruce decline indicated that there was no evidence of primary pathogen or pest damage. Recent field observations in southern Vermont indicate spruce mortality where roots are heavily infected by *Armillaria mellea,* a fungus that usually attacks stressed trees, and is associated with diebacks of other species. At this time, we do not know the extent of *Armillaria* in northern Vermont. During recent soil sampling at Camels Hump we observed a high density (~ 10 m^{-2}) of large *Lepidoptera* (genus unknown) larvae that may feed on roots. The role of these and possibly other biological agents remains to be investigated.

We regard the biogeochemical data as inconclusive, but representative of the complex nature of trying to relate acid deposition to tree mortality. More work is needed to explain differences in Al, S, N and K concentrations. Data regarding physiological stress in foliage would appear to be extremely important. We feel that no conclusions regarding the presence or absence of acid precipitation–induced stress are warranted, however we have reason to suspect that acid precipitation may be adversely affecting forest productivity.

SUMMARY

Clear and continuing changes in tree growth rate patterns have occurred in the northeastern United States that began about three decades ago. Given the suspected increase in intensity and distribution of acid precipitation during the 1950s and 1960s, it is natural to want to investigate the possibility of detrimental effects of acid precipitation on tree growth. The many pathways by which acid deposition may act, and the many combinations of abiotic and biotic stresses dictate that wide-ranging, comprehensive empirical studies are required. It is imperative to establish and test multiple working hypotheses.

REFERENCES

1. Wood, T., and F.H. Bormann. "Increases in Foliar Leaching Caused by Acidification with an Artificial Mist," *Ambio* 4(2):169–173 (1975).
2. Scherbatskoy, T., and R.M. Klein. "Response of Spruce and Birch Foliage to Leaching by Acid Mists," unpublished manuscript, University of Vermont.
3. Evans, L.S., and T.M. Curry. "Differential Responses of Plant Foliage to Simulated Acid Rain," *Am. J. Bot.* 66:953–962 (1979).
4. Tamm, C.O., and G. Wiklander. In: *Ecological Impact of Acid Precipitation,* D. Drabløs and A. Tollan, Eds. (Oslo: SNSF Project, 1980), pp. 188–189.
5. Ulrich, B. "Bodenchemische und Umwelt-Aspekte der Stabilitat von Waldokosystemen," Library, Environment Canada, Translation No. OOENV TR-2046 (1981).
6. Ulrich, B. "Die Walder in Mitteleuropa: Messergebnisse ihrer Umweltbelastung, Theo-

rie ihrer Gefahrdung, Prognose ihrer Entwicklung," Library, Environment Canada, Translation No. OOENV TR-2038 (1981).

7. Jonsson, B., and R. Sundberg. "Has Acidification by Atmospheric Pollution Caused a Growth Reduction in Swedish Forests?" Institute for Skogsproduktion Report 20 (1972).

8. Strand, L. "The Effect of Acid Precipitation on Tree Growth," in *Ecological Impact of Acid Precipitation,* D. Drabløs and A. Tollan, Eds. (Oslo: SNSF Project, 1980), pp. 64–65.

9. Johnson, A.H., T.G. Siccama, D. Wang, R.S. Turner and T.H. Barringer. "Recent Changes in Patterns of Tree Growth Rate in the New Jersey Pinelands: A Possible Effect of Acid Rain," *J. Environ. Qual.* 10(4):427–430 (1981).

10. Puckett, L.J. "Acid Rain, Air Pollution and Tree Growth in Southeastern New York," *J. Environ. Qual.* (in press).

11. Andresen, A.M., A.H. Johnson and T.G. Siccama. "Levels of Lead, Copper and Zinc in the Forest Floor in the Northeastern U.S.," *J. Environ. Qual.* 9(2):293–297 (1980).

12. Siccama, T.G. "Vegetation, Soil and Climate on the Green Mountains of Vermont," *Ecol. Monog.* 44:325–349 (1974).

13. Johnson, A.H. "Evidence of Acidification of Headwater Streams in the New Jersey Pinelands," *Science* 206:834–836 (1979).

14. Johnson, A.H. "Acidification of Headwater Streams in the New Jersey Pine Barrens," *J. Environ. Qual.* 8:383–386 (1979).

15. Cogbill, C.V., and G.E. Likens. "Acid Precipitation in the Northeastern United States," *Water Resources Res.* 10:1133–37 (1974).

16. Siccama, T.G., M. Bliss and H.W. Vogelmann. "Decline of Red Spruce in the Green Mountains of Vermont," *Bull. Torrey Bot. Club* 109:162–168 (1982).

17. Roman, J.R., and D.J. Raynal. "Effects of Acid Precipitation on Vegetation," in *Actual and Potential Effects of Acid Precipitation on a Forest Ecosystem in the Adirondack Mountains,* New York State Energy Research and Development Authority, Report ERDA 80–28 (1981), pp. 4–1 to 4–63.

18. Johnson, D.E., Oak Ridge National Laboratory, Oak Ridge, TN. Personal communication.

19. Havas, P., and S. Huttunen. In: *Effects of Acid Precipitation on Terrestrial Ecosystems,* T.C. Hutchinson and M. Havas, Eds. (New York: Plenum Press, 1978), pp. 123–132.

20. Huttunen, S., K. Laine and T. Pakonen. In: *Ecological Impact of Acid Precipitation,* D. Drabløs and A. Tollan, Eds. (Oslo: SNSF Project, 1980), pp. 168–169.

21. Huttunen, S., L. Karenlampi and K. Kolari. "Changes in Osmotic Potential and Some Related Physiological Variables in Needles of Polluted Norway Spruce," *Ann. Bot. Fennici* 18:63–71 (1981).

22. Tveite, B. In: *Ecological Impact of Acid Precipitation,* D. Drabløs and A. Tollan, Eds. (Oslo: SNSF Project, 1980), pp. 204–205.

23. J.S. Eaton, Cornell University, Ithaca, NY. Personal communication.

24. Smith, W.H. *Air Pollution and Forests* (New York: Springer Verlag, 1981).

CHAPTER 8

Use of Forest Site Index for Evaluating Terrestrial Resources at Risk from Acidic Deposition

Orie L. Loucks

A major body of literature has dealt with effects on height and diameter growth of tree species exposed to moderate concentrations of gaseous air pollutants. Many of these results have been summarized [1,2], but there still is an urgent need to consider the potential for effects on forest growth from acidic deposition. In particular, the relatively sophisticated measurement methods and models available from the field of forest mensuration should be considered for use in evaluating effects on forest growth from the alteration of forest soil chemistry following input of acidic deposition. Thus, the principal goal of this chapter is a discussion of the applicability of forest growth measurement to the detection of acidic deposition impacts on forest growth. The chapter has grown out of studies undertaken for EPA on the quantification of linked land and water system responses in watersheds [3] following addition of acidic deposition, with an emphasis on integrative measures of terrestrial and aquatic effects in the same watershed [4].

Locally focused studies have reported effects of sulfur dioxide (SO_2) gas on forests in Washington [5], Ontario [6], Virginia [7], Wisconsin [8] and Alberta [9]. These studies have shown either greatly reduced height and diameter growth, or actual mortality of the trees over a number of years, in each case attributed to local sources of SO_2 acting alone or in combination with other pollutants. Studies by Miller [1], in the San Bernardino Mountains of California, by Skelly [10], in Virginia, and by McLaughlin et al. [11], in Tennessee, have shown the more general regional response of woody species to moderate exposures of oxidants, primarily ozone (O_3). In both groups of studies, reductions have been shown in the annual rate of either height or diameter growth, with effects on height being expressed more strongly on younger stems.

Given this widespread evidence of the sensitivity of tree species to air pollutants, both locally and regionally, numerous researchers have sought to document the potential for a similar response in forest growth to the annual deposition of acidic substances in rainfall. The results to date, however, have been inconclusive

[2,12,13]. These findings may be due to imprecision in the methodologies being used to document height or diameter growth changes, they may be the result of growth having been altered already by other regionally elevated air pollutants, or the effects of acidic inputs may be neutral or even positive due to the sulfur and nitrogen nutrient content. A further objective of this chapter, therefore, is to review the background on chemical alteration of the forest environment (deposition of acidic substances and soil chemical responses), and to examine the application of forest mensurational methods in the context of a specific case study of soil nutrient stocks and site index in northern Wisconsin.

FOREST MEASUREMENT OPTIONS

A long record of forestry literature documents the diverse influences controlling height and diameter growth of trees [14–17]. This research has been cast largely in a context that recognizes the product of interest from forests (aside from recreational areas) is wood volume or its equivalent in cellulose mass. Thus, forest productivity of local study sites, watersheds or large industrial forest tracts are expressed routinely in cubic feet (or cubic meters) of annual growth per unit area (e.g., m^3-ha^{-1}-y^{-1}). This annual volume increment is controlled, in part, by the annual increase in diameter (i.e., cross-sectional or basal area), but the volume of a tree can be viewed most simply as a cone (Figure 1), and is, therefore, strongly controlled by height as well as cross-sectional area or diameter. Since, as we shall see, the diameter of the tree is more sensitive to the idiosyncracies of local competition from nearby stems, the total height of the cone (particularly for the population of cones that represents a fully stocked stand) is viewed as a more reliable measure of overall long-term forest productivity [12,13].

Of these two, the more readily measured property of trees is stem diameter. Indeed, because of the difference in density between spring wood and summer wood, most species leave a well defined record of the annual increase in diameter that can be reconstructed years later through tree-ring analysis. This record has been used [18] and is currently proposed for more extensive use in evaluating air pollutant effects on forest growth, including acidic deposition.

Problems emerge, however, because of the profound sensitivity of diameter growth to the aging of individual trees, the effects of competition from nearby stems (a density effect), annual variations in climate (particularly precipitation and temperature) and unusual local influences from shallow soils, which may not affect a young developing forest but have large effects on slowing growth in a fully developed stand. In addition, variations in responses among species add complexity to understanding the patterns of diameter responses.

Height growth of trees is linked in a general way to overall growth (including diameter growth), but the variance in mean height growth can be reduced greatly by explicit choice of "dominant" and "codominant" stems for measurement and analysis [14]. By definition, these trees have not experienced significant shading or serious competition from adjacent stems throughout their life span, thereby minimizing the density effects reflected in diameter measurements. Although a

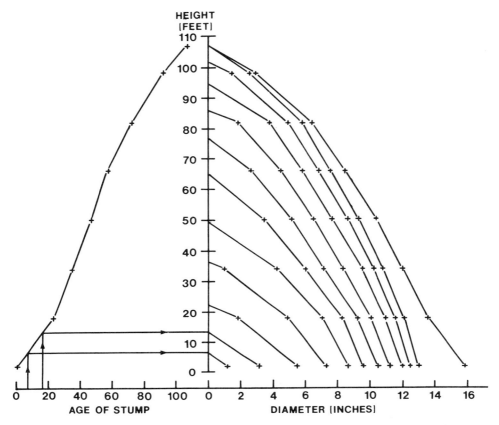

Figure 1. Tree-form analysis [14] showing the relationship between height and diameter increments in a tree as obtained by stem analysis methods. The outermost line represents outside-bark measurements; the series of concentric inner cones represent groups of annual rings in the wood which narrow rapidly at the base in later years.

general relationship exists between height growth and fluctuations in climate, much of the height increment occurs as a burst of growth in the spring derived in part from the previous year's carbohydrate accumulation and partly from the current year's growing conditions.

Most importantly, the gradual development of the height of an entire stand is a property of the whole forest, as well as of the individual dominant stems. Therefore, height represents a condition of the forest as it responds to its environment, including any alterations of that environment by chemical pollutants. For example, the effect of oxidants on both height and diameter of ponderosa pine is shown in Figure 2 [1]. Thus, to evaluate the alteration of forest productivity over some period of time, as might be the case with acidic deposition, one should use a stand property that averages over a multiyear growing period and can capture the average height response of the population of stems controlling total volume. Because a stand can be defined as fully stocked at either high or low density

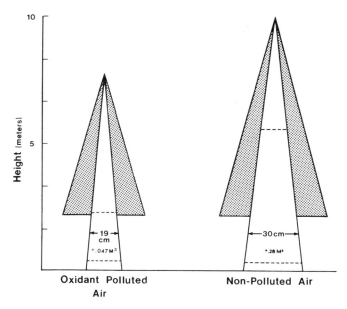

Figure 2. Calculated average growth of 30-year-old, 15-cm ponderosa pines at San Bernardino National Forest, California, in polluted and nonpolluted air based on radial growth samples from 1941–1971 and 1910–1940 [1]. The asterisk indicates wood volume in log with 15-cm top (minimum merchantable diameter).

(i.e., low or high average diameter), the plastic measure represented by diameter growth is difficult to use as a long-term characterizing property of the stand.

Consideration of model characteristics needed for projecting future responses also indicates the desirability of working with a measurement property with a background of predictive uses. The forestry literature has developed such a measure, known primarily as "forest site index" (FSI) [14,15]. This measure expresses stand height growth for a given combination of soil and species over a standard time interval such that in the absence of human alteration consistent values are obtained from one forest rotation to the next. Site index is most commonly expressed as the average height of dominant and codominant trees at an age base of 25–50 y (in the East) or 100 y (in the West). Measurements taken on trees younger or much older than this age base can be converted to the standard base through the use of established curves or tables, available for most species [14,16,17].

FSI has been used for almost a century in Europe and North America as a standard measure of both current and potential forest productivity. The wide agreement on its use derives, in part, from the facility with which the height of trees, multiplied through full stocking or standard basal area, converts to cubic volumes or dollar value of wood products. Decades of research on growth rates and harvestable yields in relation to height, stem density and basal area (average diameter) are summarized in standard yield tables [16].

Quantifying pollutant impacts on a resource such as a forest stand requires one to distinguish among properties of the system that have varying degrees of sensitivity to pollutant exposures. The observer must differentiate between two general groups of measures: "indicators," individual high-resolution response measures of components within a complex system (i.e., a chemical nutrient stock), and "integrators," measures that reflect the combined action of several aspects of the environment governing a more aggregated response of the components [4]. FSI is representative of the latter. Both types of measures are necessary to present fully the combined effects of pollutants on a complex system, such as a forest exposed to the many variables present in the environment. Annual height and diameter growth of forests, which represent year-long responses to both air and soil chemistry, represent an annually integrative measure of the responses to these conditions, but FSI represents a longer-term measure, and one that is predictive with respect to the next generation of forests on that site.

BACKGROUND ON ACIDIC INPUTS AND FOREST EFFECTS

Studies of the effects of acidic deposition on forest growth have been underway for some years in Sweden and Norway and have been summarized [19–22]. The early work in Sweden by Jonsson and Sundberg [19] used tree-ring analysis methods, and the finding of possible effects was couched in considerable uncertainty. A later study was not able to support the original findings of effects from acidic deposition, but more recent research in Sweden, shows that certain very sensitive soil conditions, negative effects on tree growth can be demonstrated [23]. On naturally acid soils and on fertile, well-buffered soils no effects are expected and none are being found.

Experimental approaches for detecting effects of acidic deposition on tree growth have been undertaken in Norway [12,20,21]. These have led to the conclusion that over the period of a few years, the initial beneficial effects (on some soils) from the nitrogen and sulfur content in acidic deposition can be reversed over a period of time. In addition, the negative influence apparently due to the loss of soil cations (and possibly the toxicity of moderately elevated aluminum concentrations) combines to reduce the growth of trees [21]. The more general findings by Abrahamsen [12] suggest the effects from acidic input may be negative more frequently on soils of only moderately low nutrient status but potential magnesium or potassium deficiency. Where nitrogen and sulfur are the principal nutrient deficiencies, however, the response is likely to be positive in the short term, but negative in the long run. Abrahamsen gives no specific criteria as to how widespread each of these soil deficiency types may be in Norwegian forests.

Studies in Germany [24] have led to the conclusion that mobilization of aluminum (under situations of unusually high H^+ concentration) has produced negative effects on Norway spruce fine roots such that reduced growth and associated disease symptoms (extending to mortality) are being observed. Similar effects on diameter growth in association with fluctuations in rainfall pH and a downtrend in stream pH have been reported by Johnson et al. for pitch pine in New Jersey [18]. Other studies reported by Johnson et al. (Chapter 7) show altered growth rates of red spruce in Vermont. Toxicity from free ionic aluminum (Al^{3+}) mobilized by acidic deposition, particularly during flushing episodes such as described by Johnson et al. [25], appears to be the most plausible explanation.

The results of several studies in Scandinavia, particularly those of Bjor and Teigen [22], suggest that estimates of the magnitude of cation removal from the surface layers of soil due to measured acid inputs may be possible. For the case of the shallow soils studied in Norway, a total stock of calcium + magnesium + potassium in the surface layers (in meq-m^{-2} to a 35-cm depth) can be summarized for both "before" and "after" treatment. The differences, representing experimentally induced time treatments of 0.4–280 y, indicate a loss of up to 50% of the Ca and a smaller portion of the Mg and K over the full period, even after allowing for the intervening release of weathering products. The total quantities leached during the experiment are larger than the initial nutrient stocks.

An analysis of the prospective effects on forest growth from acidic deposition must take into consideration and try to quantify the following four pathways by which effects can be expressed:

1.   magnitude of growth reduction by exposure to oxidants;
2.   direct foliar (carbon uptake) effects of acidity and other constituents in acid-altered precipitation;
3.   effects on tree growth that may be attributable to a reduction in the stock of the nutrient cations in the soil or associated alterations of the cycling of nutrients such as N, P, K and S; and
4.   extent to which episodic mobilization of toxic forms of aluminum in the soil solution can produce effects on tree roots and, therefore, on aboveground growth responses.

Indeed all four of these factors could operate simultaneously, and a potential exists for synergistic effects.

CASE STUDY FROM WISCONSIN

A data base within which to examine the utility of forest mensurational methods for analysis of pollutant-related tree-growth effects exists from the mid-1960s for quaking aspen in northern Wisconsin [17,26]. A portion of these data has been adapted specifically for obtaining aspen site index estimates solely on the basis of physiographic, soil and nutrient stock data for these sites (Table I). Simple regression models were developed from the site index and soil chemical data, with the results shown in Table II. The low (to negative) correlations reflect both the low nutrient demand of aspen and, in the case of phosphorus, an apparent interaction between water-holding characteristics and soil phosphorus (dry soils with slightly higher P, soils with a water table at depth but low P). Nitrogen levels showed no correlation with growth rates.

Table I. Mean and Range of Site Index and Independent Variables for 32 Quaking Aspen Stands [17]

Variable	Mean	Range
Site Index	76.5	65–88
Sand, B2 (%)	52.1	2.0–94.0
Silt, B2 (%)	30.0	2.0–80.0
Clay, B2 (%)	15.3	6.0–30.0
AWC (cm of Water)		
12-in. Depth	2.48	1.4–4.4
60-in. Depth	10.9	5.1–16.7
Water Table Depth (in.)	101.8	40–120+
Exposure	3.2	1–5
Ca (lb/ac to 12 in.)	1,044.6	303.3–3,027.6
Mg (lb/ac to 12 in.)	281.1	60.9–570.1
K (lb/ac to 12 in.)	157.4	99.8–257.0
P (lb/ac to 12 in.)	82.7	16.6–318.3
Ca (lb/ac to 60 in.)	5,946.4	1,505.5–15,967.8
Mg (lb/ac to 60 in.)	2,001.3	227.2–6,203.8
K (lb/ac to 60 in.)	854.9	419.5–1,666.9
P (lb/ac to 60 in.)	485.0	45.6–1,060.1

Table II. Correlations Between Site Index and Nutrient Levels for 30 Quaking Aspen Stands in Northern Wisconsin[a] [25]

Nutrient (lb/ac)	Correlation Coefficients			
	12-in. Depth	24 in. Depth	36-in. Depth	60-in. Depth
P	−0.474[b]	−0.454[b]	−0.514[b]	−0.557[b]
K	0.277[c]	0.420[c]	0.440[b]	0.469[b]
Ca	−0.071	0.109	0.240	0.311[c]
Mg	0.087	0.344[c]	0.392[c]	0.392[c]

[a] Only stands on sites with water tables deeper than 120 in. are included.
[b] Significant at 0.01 level.
[c] Significant at 0.05 level.

In addition, multiple regression models were developed and tested; the principal model was [17]:

$$\text{Aspen SI}_{(\text{age } 50)} = 73.48 + 3.85\text{AWC} - 0.09\text{WTD} + 1.64\text{Exp} + 0.0012\text{Mg}$$

where   AWC = available water capacity (cm of water), to 60 inches
            WTD = water table depth (in.)
              Mg = magnesium nutrient stock (lb/ac)
            Exp = stand exposure expressed in a nominal scale

Table III.   Initial Magnitude and Posttreatment Quantities of Three Cation Nutrient Stocks in a Norwegian Study and a Comparison with Stocks in Northern Wisconsin [17]

	Ca	Mg	K
Norwegian Study			
Initial			
(meq-m^{-2} to 35 cm)	125	20	28
Posttreatment			
pH 4 (9 years)			
meq-m^{-2} to 35 cm	98	14	25
% lost	22	30	11
pH 3 (90 years)			
meq-m^{-2}	14	2	24
% lost	89	90	14
Wisconsin Study			
Initial (lb/ac to 12 in.)	300–3,000 (1,050)	60–570 (280)	100–250 (155)
Posttreatment	1,680–16,800 (5,880)	558–5,301 (2,604)	287–717 (445)
(meq-m^{-2} to 12 in.)			

This equation, however, accounts for only 68% of the variation in site index. The variability of the data base for these models is illustrated by the means and range of major site properties shown in Table I. Ca and K showed small positive regression coefficients, but they correlate with Mg and only one of the nutrients is needed in the regression.

The Bjor and Teigen results [22] suggest the possibility, however, that at some future time it may be possible to estimate experimentally the long-term cation removals from poorly buffered soils in a region such as northern Wisconsin. With a sustained application of acidic materials (applied at pH 3 and 4), their data (Table III) suggest that over a period of 2–3 decades, perhaps 40–70% of the calcium and magnesium nutrient stocks could be leached from the surface rooting zone of their very low-nutrient soils. The corresponding stocks of these three elements in northern Wisconsin (to a depth of 12 in.) are summarized in the lower part of Table III. Even the poorest soils in northern Wisconsin show 10–20 times the Norwegian nutrient stocks, however, and average sites are about fivefold higher still. Since very small residual cation stocks tend to resist leaching and are retained as a continuing low percentage base saturation, the low-nutrient sites in Wisconsin could show similar cation losses for the same relatively strong acidic inputs, but the percentage reduction would be lower due to the higher initial stocks. For an illustrative computation, and given the pH 4.5 rainfall of northeastern Wisconsin [4], one could say that on the order of 10–30% of the Ca and Mg stock might be leached from the low-nutrient sites in a 25-y period, and 2–5% from the intermediate sites. The acid neutralizing capacity of the "high nutrient" sites appears to be sufficient to neutralize all acid inputs without appreciable percentage reductions in Ca and Mg stocks.

Table IV. Example Regression[a] Solutions for Effects of Mg Leaching on Aspen Site Index in Northern Wisconsin to 1990

	Mg Stock × Mg r Coeff.	Calculated Aspen SI (ft)	ΔFSI 1965–1990 (ft)
Case 1[b]			
1965 Mg Effect	220[c] × 0.0012	73.8	
Possible 1990 Mg Effect	220 × 0.75[d] × 0.0012	73.7	0.1
Case 2[e]			
1965 Mg Effect	2000[c] × 0.0012	75.9	
Possible 1990 Mg Effect	2000 × 0.97[f] × 0.0012	75.8	0.1

[a] Regression abscissa = 73.5 ft.
[b] 25% loss of Mg over 25 y on "low-nutrient" sites.
[c] Regression model was calibrated to total nutrient stocks (lb/ac) to 60 in., rather than the 12-in. value shown in Table III.
[d] Assumes, for illustration, a 25% leaching of the Mg stock over 25 y.
[e] 15% loss of Mg over 25 y on "intermediate" sites.
[f] Assumes, for illustration, a 3% leaching of the Mg stock over 25 y.

These calculations of possible cation losses, together with the prealteration regression model that includes a cation component, suggest a procedure by which one might estimate effects from this one aspect of pollutant loadings (cation stripping) on future forest growth rates. In effect, such computations represent a hypothesis to be tested against future field observations over several decades. An example computation is shown in Table IV, using Mg alone (the only cation whose influence on aspen site index in northern Wisconsin was significant). The regression model for trembling aspen is shown for the effect of Mg in the baseline year (1965), as well as for the year 1990 with a reduced stock of Mg. Given the modest amount of the variation in site index accounted for by the model, no significance can be attached to the estimate of 0.1 ft reduction at 50 y, but the methodology is illustrated.

At the low range of aspen site index, factors other than nutrient status appear to dominate growth, possibly because the multiple regression form without transformations tends to linearize the response over the entire observed range. With further analysis, the effects of Mg and Ca limitation on site index in the low nutrient range could be much larger. The effect of a shift of 0.1 ft in site index calculated for 1990 under either low or intermediate nutrient stocks can be summarized as follows:

- From aspen yield tables [27], a FSI of 80_{50} (i.e., 80 ft at 50 y) gives a mean annual growth of 70.6 ft³/ac/y.
- For aspen FSI 70_{50}, mean annual average growth is 62.15 ft³/ac/y.

Thus, for the above alteration in site index, 0.1 ft at 50 y by 1990, for "low" and "intermediate" sites, the change in total growth can now also be calculated. Assuming a linear interpolation in the difference in harvest at 50 y for SI

80 vs SI 70 (3531 vs 3103 ft³, respectively), the change induced by a loss of 0.1 ft in height would be 5 ft³.

Since other interactions also may come into play when a single factor is manipulated without transformation at the data limits of a multiple regression model, and these could influence the results, the effects of Ca and Mg depletion from low-nutrient sites could be larger than indicated here. Therefore, these results illustrate a methodology and suggest that all aspects be evaluated fully during the course of further studies. The small indicated effect illustrated here could be much larger on the lowest-nutrient sites even for aspen, and larger for other species or still other regions with nutrient stocks more similar to those in Norway.

DISCUSSION

The above computations have focused on only one of the four principal pollutant effect pathways cited earlier in the chapter, but they serve to illustrate a possible predictive mode within which forest growth effects may need to be considered. Oxidant effects already can be expressed as wood volume reductions by modeling the alteration of foliage condition, and the direct effects of gaseous pollutants are becoming well documented from studies in local impact areas.

Of more concern is that, to date, all estimation procedures have had to neglect the potential effects of aluminum toxicity. The most important experimental data presently available for tree species are those of McCormick and Steiner [28], from a study showing that one variety of hybrid poplar was sensitive to moderate aluminum concentrations. The other species tested, at least under the experimental approach used (i.e., in aqueous solutions, rather than a medium utilizing fungal relationships between roots and soil), showed little effect from aluminum over the range of concentrations likely to be observed even in acid soils.

An important aspect of this question is the current indication of Al^{3+} concentration likely to be expressed in the rooting zone of soils experiencing acidic deposition. No good measurement methodology is available for determining these concentrations during the rapidly changing conditions of an acid flushing event. However, computational procedures developed by Johnson et al. [25] are available which use the readily measured H^+ concentrations. This methodology shows that in headwater streams during flushing events, in which a portion of the inflowing water has been in brief contact with surface layers of the soil, aluminum concentrations approaching 0.25 mg-L⁻¹ are being observed. Given the large (but relatively uncertain) dilution for Al^{3+} from soil solution concentrations to Al^{3+} in the stream water, a significantly higher concentration around the fine roots of trees in the surface layers is likely. Although determining precisely the significance of brief exposures of roots of trees in the natural environments is a critical research need, these data suggest Al^{3+} concentrations in the surface soil layers will be in the range of a few parts per million for short periods, a level that would be significant for the hybrid poplar studied by McCormick and Steiner [28].

Finally, the question of using the more readily available diameter measurements from tree ring analysis to reconstruct a stand response in terms of basal area, height growth, FSI or total forest productivity also must be addressed. Many of the tree-ring methodologies use a sufficiently large number of stems for analysis that a concept such as "the tree of average diameter" can be defined and used. By a methodology such as stem analysis (see Figure 1), it is possible to reproduce the height development of a tree of "average diameter" and develop an algorithm for the change in rate of height growth (and of site index) associated with the pollutant-induced changes in diameter growth.

Several other considerations, however, may limit the FSI approach. One relates to the widespread use of all-aged management in forests, and the great difficulty in obtaining freely growing stems for measurement under these conditions. A second consideration stems from the basic assumption that the shape of the height/age response over time is relatively uniform over various sites. Although this is usually the case, it does not hold for all soils and produces variance that cannot be incorporated in most models predicting growth rate. To reduce differences due to these limitations, standard site index measurement methods must be used and interpreted in conjunction with some knowledge of the forest age-class distributions and specific soil types in the regions involved.

CONCLUSIONS

The goals here have been primarily an examination of a methodology that may be needed if the effects on forest growth from acidic deposition are to be quantified and evaluated in detail. Effects on growth from cation stripping alone, or any other single effect pathway, may never be found to be a controlling factor, but methods will be needed that consider all potentially significant mechanisms.

The effects of acid precipitation on a watershed/lake system are diverse and intricately balanced, especially when a mature forest ecosystem holding a large nutrient stock is present. As outlined above, forest mensurational methods, particularly site index, appear to offer an "integrative" tool for expressing the effects on or changes in forest productivity over many decades due to changes in pollutant loads over such periods. Use of the site index as an integrator of acid deposition effects on growth (via nutrient stripping or aluminum toxicity) ultimately has the principal advantage of being readily converted to harvestable product or other economic terms for cost/benefit analysis.

REFERENCES

1. Miller, P.R., Ed. *Photochemical Oxidant Air Pollutant Effects on a Mixed Conifer Forest Ecosystem—A Progress Report,* EPA-600/3–77–104, U.S. Environmental Protection Agency, Corvallis, OR (1977).
2. Smith, W.H. *Air Pollution and Forests* (New York: Springer-Verlag New York, Inc., 1981).

3. Loucks, O.L. "Models Linking Land-Water Interactions Around Lake Wingra, Wisconsin," in *Land-Water Interactions,* A.D. Hasler and J. Olson, Eds. (New York: Springer-Verlag, 1975), pp. 53–63.
4. Loucks, O.L., R.W. Usher, D. Rapport, W. Swanson and R.W. Miller. "Assessment of Sensitivity Measures for Evaluating Resources at Risk from Atmospheric Pollutant Deposition," Final Report to the U.S. Environmental Protection Agency, ERL, Duluth, MN (1981).
5. Scheffer, T.C., and G.C. Hedgcock. "Injury to Northwestern Forest Trees by Sulfur Dioxide from Smelters," Tech. Bull. No. 1117, USDA Forest Service, Washington, DC (1955).
6. Linzon, S.N. "Economic Effects of Sulfur Dioxide on Forest Growth," *J. Air Poll. Control Assoc.* 21(2):81–86 (1971).
7. Stone, L.L., and J.M. Skelly. "The Growth of Two Forest Tree Species Adjacent to a Periodic Source of Air Pollution," *Phytopathology* 64:773–778 (1974).
8. Karnosky, D.F. "Changes in Southern Wisconsin White Pine Stands Related to Sulfur Dioxide on Forest Growth," *Symposium on Effects of Air Pollutants on Mediterranean and Temperate Forest Ecosystems,* P.R. Miller, Ed., General Technical Report PSW-43, USDA Forest Service, Berkeley, CA (1980), pp. 238–239.
9. Legge, A.H. "Primary Productivity, Sulfur Dioxide, and the Forest Ecosystem: An Overview of a Case Study," in *Proceedings of a Symposium on Effects of Air Pollutants on Mediterranean and Temperate Forest Ecosystems,* P.R. Miller, Ed. General Technical Report PSW-43, USDA Forest Service, Berkeley, CA (1980), pp. 51–62.
10. Skelly, J.M. "Photochemical Oxidant Impact on Mediterranean and Temperate Forest Ecosystems: Real and Potential Effects," in *Proceedings of Symposium on Effects of Air Pollutants on Mediterranean and Temperate Forest Ecosystems,* P.R. Miller, Ed., General Technical Report PSW-43, USDA Forest Service, Berkeley, CA (1980), pp. 38–50.
11. McLaughlin, S.B., R.K. McConathy and D. Duvick. "Effects of Chronic Air Pollution Stress on Allocation of Photosynthate by White Pine," in *Proceedings of Symposium on Effects of Air Pollutants on Mediterranean and Temperate Forest Ecosystems,* P.R. Miller, Ed., General Technical Report PSW-43, USDA Forest Service, Berkeley, CA (1980), pp. 244–245.
12. Abrahamsen, G. "Acid Precipitation, Plant Nutrients and Forest Growth," in *Ecological Impact of Acid Precipitation,* D. Drabløs and A. Tollan, Eds. (Oslo: SNSF Project, 1980), pp. 58–63.
13. Overrein, L.N., H.M. Seip and A. Tollan. "Acid Precipitation Effects on Forest and Fish; Final Report of the SNSF Project, 1972–1980," Oslo, Norway (1980).
14. Bruce, D., and F.X. Schumacher. *Forest Mensuration* (New York: McGraw-Hill, 1950).
15. Spurr, S.H. *Forestry Inventory* (New York: Roland Press Company, 1952), pp. 299–320.
16. Forbes, D., Ed. *Forestry Handbook.* (New York: Roland Press Company, 1955).
17. Fralish, J., and O. Loucks. "Site Quality Evaluation Models for Aspen (*Populus tremuloides* Michx.) in Wisconsin," *Can. J. Forest Res.* 5(4):523–528.
18. Johnson, A.H., T.G. Siccama, D. Wang, R.S. Turner and T.H. Barringer. "Recent Changes in Patterns of Tree Growth Rate in the New Jersey Pinelands: A Possible Effect of Acid Rain," *J. Environ. Qual.* 10(4):427–430 (1981).
19. Jonsson, B., and R. Sundberg. "Has the Acidification by Atmospheric Pollution Caused a Growth Reduction in Swedish Forests?" Report 20, Inst. For Skogsproduktion (1972).
20. Abrahamsen, G. "Impact of Atmospheric Sulfur Deposition on Forest Ecosystems,"

Atmospheric Sulfur Deposition, D.S. Shriner, C.R. Richmond and S.E. Lindberg, Eds. (Ann Arbor, MI: Ann Arbor Science Publishers, Inc., 1980), pp. 397–415.

21. Tveite, B. "Effects of Acid Precipitation on Soil and Forest. 9. Tree Growth in Field Experiments," in *Ecological Impact of Acid Precipitation,* D. Drabløs and A. Tollan, Eds. (Oslo: SNSF Project, 1980), pp. 206–207.

22. Bjor, K., and O. Teigen. "Effects of Acid Precipitation on Soil and Forest. 6. Lysimeter Experiment in Greenhouse," in *Ecological Impact of Acid Precipitation,* D. Drabløs and A. Tollan, Eds. (Oslo: SNSF Project, 1980), pp. 200–201.

23. Anderson, F. Personal Communication.

24. Ulrich, B., R. Mayer and P.K. Khanna. "Chemical Changes Due to Acid Precipitation in a Loess-Derived Soil in Central Europe," *Soil Sci.* 130:193–199 (1980).

25. Johnson, N.M., C.T. Driscoll, J.S. Eaton, G.E. Likens and W.H. McDowell. "Acid Rain, Dissolved Aluminum and Chemical Weathering at the Hubbard Brook Experimental Forest, New Hampshire," *Geochim. Cosmochim. Acta* 45:1421–1437 (1981).

26. Fralish, J.S. "Site Indices and Rates of Conversion in Northern Wisconsin Aspen," PhD Dissertation, University of Wisconsin, Madison, WI (1969).

27. Schlaegel, B.E. "Growth and Yield of Managed Stands," in *Aspen Symposium Proceedings,* General Technical Report NC-1, USDA Forest Service, North Central Forest Experiment Station, St. Paul, MN (1972), pp. 109–112.

28. McCormick, L.H., and K.C. Steiner. "Variation in Aluminum Tolerance Among Six Genera of Trees," *Forest Sci.* 24:565–568 (1978).

INDEX

Charles Seale-Hayne Library
University of Plymouth
(01752) 588 588
LibraryandITenquiries@plymouth.ac.uk